D0553928

CEREAL PESTS AND DISEASES

CEREAL PESTS AND DISEASES

By

R. GAIR
B Sc., FI Biol.

J. E. E. JENKINS
B Sc., MS (Cornell), MI Biol.

and

E. LESTER
B Sc., FI Biol.

Pest sections revised
by
PETER BASSETT
B Sc., MI Biol.

FARMING PRESS LIMITED
WHARFEDALE ROAD, IPSWICH, SUFFOLK

First published 1972
Second impression 1976
(*with amendments*)
Second (*revised*) *edition* 1978
Third edition 1983
Fourth edition 1987

British Library Cataloguing in Publication Data

Gair, R.
Cereal pests and diseases.—4th ed.
1. Grain—Diseases and pests—Great
Britain
I. Title II. Jenkins, J. E. E. III. Lester,
E. IV. Bassett, Peter
633.1′049′0941 SB608.G6

ISBN 0–85236–164–5

Cover design by Hannah Berridge

Cover photographs feature barley loose smut and *septoria tritici* leaf spot of wheat. Those of bird-cherry aphid and wheat bulb-fly larva are by Duncan Smith.

Acknowledgements

WE ARE grateful to a number of people and organisations for providing us with photographs. They include *Arable Farming*; the Boots Company Ltd; Farming Press; the Department of Cryptogamic Botany, Long Ashton Research Station, University of Manchester; the National Institute of Agricultural Botany; Rothamsted Experimental Station; RHM the Lord Rank Research Centre; the Shell Chemical Company Ltd; and the Welsh Plant Breeding Station. Several photographs (Crown copyright) are reproduced by permission of the Controller of Her Majesty's Stationery Office.

R.G.
J.E.E.J.
E.L.

This book is set in 10pt on 11pt Times and is printed in Great Britain on Fineblade Cartridge paper by Page Bros (Norwich) Ltd., Norwich

CONTENTS

ILLUSTRATIONS

PAGE

COLOUR PLATES

PLATES

FOREWORD

By F. R. HORNE

CBE, MA, FRAgS, FIBiol, NDA, NDD (Hons)

THE UNDOUBTED potential of the land and of improved new varieties is not always achieved in practice. This is a common experience, even in Britain where the yields of wheat, barley and oats regularly come near to the top for Europe and indeed for the world. Plant diseases are often major factors in limiting the development of our crop plants, thus causing reduction in yield and in quality of grain, and variability from one season to the next.

It is most uncommon to have the wide ranging experience of three such outstanding authorities as Robert Gair, John Jenkins and Joe Lester brought together into one volume, which is illustrated with excellent photographs—a great help in recognising the diseases.

This book is written primarily for growers and others who have a practical interest in our cereal crops; there are also sections on maize and on rye. A great deal of interesting and valuable information is given on cultural practices as well as on chemical control and the value of improved varieties.

I hope the publication will encourage others to make known their experiences in the practice of disease control. It is by pooling our knowledge of different regional conditions and systems of management that we can greatly influence the dimensions of the problems and help to overcome them.

Coming at a time when disease such as yellow rust, mildew and *Rhynchosporium* are making their presence felt on our extended acreages of cereals, this very readable and interesting book can be recommended to all who are concerned with cereal production.

September, 1972 F. R. HORNE

PREFACE
TO THE FIRST EDITION

THIS BOOK aims at providing farmers and other agriculturalists with a guide to the many pests, diseases and disorders which afflict cereals growing in Britain. It gives an account of the damage, symptoms and life histories of pests and diseases, potential and (where possible) actual crop losses, and the measures which can be taken to combat attacks.

For quick access to the information the contents are arranged so that for each crop—wheat, barley, oats, rye and maize—there is one chapter dealing with the pests and another with the diseases which attack them. A particular pest or disease often affects more than one cereal, in which case it is given more attention in the chapter relating to the crop with which it is most commonly associated. It is then mentioned or briefly discussed under the other crops, the reader referred to the fuller account in each case. Thus repetition has been kept to a minimum consistent with the objective of providing a reasonably comprehensive account of the pests and diseases of each crop in a separate chapter.

Each chapter on pests begins with a guide based on plant damage symptoms. The various pests are then described in the following order—first insects, then millepedes, slugs, nematodes (eelworms), birds and lastly mammals.

Each chapter on diseases begins with a guide to diseases and disorders arranged according to the part of the plant damaged or that showing symptoms—seedlings, roots and stem bases, leaves and stems, ears and grain. The accounts of disease follow in the same order as the guide except that where a disease or disorder affects more than one part of the plant, an account of all its phases is given in one place. Thus mildew and *septoria* diseases are included with leaf and stem diseases although both may also attack the ears.

This use of guides to each chapter is adopted because it is assumed that most readers would initially be concerned with a particular crop and its problems. Readers concerned with a particular pest or disease may refer to the index at the end of the book.

ROBERT GAIR
JOHN JENKINS
ERIC LESTER

June, 1972

PREFACE
TO THE FOURTH EDITION

DEVELOPMENTS SINCE 1982, more particularly in some diseases, have necessitated re-writing some sections. Elsewhere the information has been updated, especially that in relation to chemical control measures and the problems associated with fungicide resistance which has recently become a significant factor in disease control.

JOHN JENKINS
PETER BASSETT

August, 1986

Chapter 1

INTRODUCTION

CEREALS HOLD a dominant position among arable crops in the United Kingdom with about 4 million hectares representing three-quarters of the arable area devoted to these crops annually. This level of intensification was reached by a steadily increasing cereal area beginning about the mid 1950s. The intensification reached a plateau by the late 1960s, and by 1970 serious losses from pests and diseases had been recognised.

The risks from diseases were exacerbated by the limited number of varieties of wheat and barley grown, two varieties in each case occupying two-thirds and one-half the area respectively. By the 1980s, the range of dominant varieties was certainly greater, although until recently one variety still accounted for 35 per cent of the UK area devoted to winter wheat, and in Scotland spring barley continued to be dominated by a single variety. Winter barley, it is estimated, increased from about 15 per cent of the barley area in 1977 to about 50 per cent in the 1980s (and considerably more in some areas), with most of the area sown to two or three varieties.

Cereal yields have increased markedly, especially on the better soils, as farmers have adopted a more systematic approach to growing. Much effort has been devoted to the problems of putting complementary techniques into systems to best advantage for a particular farming area. Cereal disease control has played a prominent part with the advent in the recent decade or so of more durable forms of resistance, the adoption of minimum disease resistance standards, the concept of varietal diversifcation and the availability of effective fungicides.

With the exception of cereal aphids, pests do not often cause widespread damage on the scale of such diseases as barley mildew or take-all of wheat, but they are important in particular situations and would be even more damaging were it not for the efficient use of insecticides.

SYMPTOMS

Plants may be damaged by pests such as rabbits, wireworms, aphids; by infectious diseases such as mildews and rusts, and by non-infectious diseases such as those caused by nutritional deficiencies and weather damage. We recognise these disorders by the symptoms which appear on the plant. Sometimes the causal agent is present at the exact place on the plant which develops symptoms. Thus in the case of the wheat bulb fly, the grub which kills the shoot may be found within it; and in the case of rusts, the premature death of the leaf is associated with the presence of many rust pustules on the same leaf. In other cases, symptoms and causal agents are less directly associated; for example the take-all fungus causes a root rot, but a more obvious symptom may be seen in the above-ground parts where the plant is stunted and the ear is a poorly filled 'whitehead'. In the cases of virus diseases and nutritional disorders, the causal agent may be present (or absent) throughout the plant, and the symptoms may be localised or general.

Since symptoms are mainly the reaction of a plant to a pest or disease, and as the plant is endowed with limited means of expression, it is not surprising that different pests and diseases often produce very similar symptoms in the host plant. Because of this, correct identification of the causal agent is essential before control measures can be considered, and this often requires the assistance of a pathologist or entomologist with laboratory facilities.

SOURCES AND SPREAD

Most of the infectious diseases of plants are caused by fungi. Some of these are soil-borne and tend to spread only slowly within a field; some are seed-borne, either as contaminants on the surface or as infections within the seed; some are spread by airborne spores. This last group, which includes the important leaf diseases, is the one often associated with widespread damage. Rusts and mildews can spread rapidly from field to field, sometimes over considerable distances. Given favourable conditions for spread and a large area of susceptible varieties, these epidemic diseases can cause severe damage on a country-wide scale.

Many cereal pests are primarily associated with grassland and those are referred to later as 'ley pests'. They transfer or migrate to cereals when their natural grass host is destroyed, for example by ploughing. Because of this there is a clear trend for many pests to assume less importance in intensive cereal systems than in ley farming or alternate husbandry systems. There are of course exceptions to such a general rule, as for example cereal cyst nematodes, which may multiply with increasing frequency of cereal cropping, and cereal aphids, which invade in spring and summer from outside sources quite independently of the cropping system.

These aphids may be carriers of barley yellow dwarf virus if, before arrival, they have fed on infected cereals or grasses. Particularly damaging attacks of this disease can occur when cereal aphids capable of transmitting virulent forms of the virus migrate from infected grasses to newly emerging autumn-sown cereals.

Nutritional disorders do not of course spread since they are not infectious. Their occurrence is often closely correlated with soil type, as for example, manganese deficiency which most frequently occurs on organic soils of high pH. One form of copper deficiency is closely associated with chalk downland soils of the Icknield with potash deficiency. Other non-infectious disorders are often associated with weather damage, especially frost.

HOST RANGE AND SPECIALISATION

Nearly all the fungi which affect cereals have a limited host range. Some such as the take-all fungus may affect wheat, barley and some grasses; some such as *Septoria* (which causes leaf spots and glume blotch) affect a similar host range, but the forms that attack a particular host tend to be less damaging to other hosts.

Another group of fungi, which are highly specialised in respect of their host range, includes the powdery mildews and the rusts. A particular mildew or rust species may affect several hosts, and the fungus on one host is indistinguishable by normal means from that on another. However, specialised forms of these fungi exist, each able to attack only one crop species, so that the form on wheat cannot attack barley, oats or other crops, and the forms on these hosts will not affect wheat. Within each of the specialised forms able to infect a particular host there are numerous so-called 'physiologic races' or 'strains' distinguishable one from another only by the range of varieties they are able to infect.

This specificity of fungus or strain of fungus to the crop is of the greatest practical significance when considering crop rotation, sources of infection and their relative importance and the development of epidemics. Thus, barley will not contract mildew or rust from volunteer plants of wheat nor from adjacent wheat crops and vice versa. The unspecialised pathogen causing take-all, however, can infect both crops and will pass from one to the other easily.

Specialisation also occurs with certain pests. Races of cereal cyst nematode, often called biotypes or pathotypes, cannot be distinguished from each other by shape or size but differ in their ability to develop into females (cysts) or host plants. Stem nematode (*Ditylenchus dispaci*) also exists in a number of races, each of which has its own particular host range, sometimes extending to many crops other than cereals.

FACTORS AFFECTING PESTS AND DISEASES

Many factors influence the occurrence and severity of pests and diseases and will influence different ones in different ways. Some have large effects and some small and the combined effects of interaction between the many pests and diseases which can affect crops at the same time are complex. The following notes are intended only to give an indication of the likely effects of some of the more important factors.

Weather

Rainy summers encourage slugs, rust, take-all, brown foot rot and ear blight, loose smut, septoria leaf spots and barley leaf blotch.

Warm dry summers lead to greater damage from mildew (which is, however, inhibited by very warm weather) but overall result in lower levels of disease and of slug numbers and activity. Spread of loose smut will be restricted. There is a risk of drought, especially on lighter soils and in spring-sown crops. Such conditions will seriously increase damage caused by most diseases and may predispose plants to late attacks of brown foot-fot.

Leatherjackets are very susceptible to drought when the young larvae are newly hatched and numbers fall dramatically if rainfall in autumn months is much below average.

A wet spring encourages invasion of cereal roots by the cereal cyst nematode and also favours eyespot infection. May yellows, due to temporary nitrogen shortage, may be common and consequently the damage caused by take-all, which is exacerbated by nitrogen deficiency, may be disproportionately high compared with the level of infection.

After a mild, open winter and spring, cereal aphids can be expected in abundance and yellow rust may occur early. Take-all attacks may be severe on winter wheat or barley, but conversely may be less in evidence on spring-sown crops since these conditions hasten the disappearance of the fungus in uncropped land.

Soil Type

On light-textured soils there is a risk of greater damage from cereal cyst nematode and from rhizoctonia root rot. Take-all can develop rapidly on these soils especially if they are alkaline.

On heavy soils slugs are a notable risk, while among the diseases eyespot and brown foot rot may be more common. Take-all is slower to develop and is usually less damaging on such soils.

Cultivations

Various cultivation systems that avoid soil inversion have been developed and are often used in cereal growing. They range from direct drilling to the use of discs and chisel ploughs but all leave most or all of the debris from a previous cereal crop on the soil surface. This might be expected to increase the carry-over of debris-borne diseases such as eyespot, septoria in wheat and rhynchosporium and net blotch in barley, and as a consequence to increase the severity of disease in the subsequent crop. In practice leaf diseases are occasionally more severe than when conventional mouldboard ploughing is practised, but in the majority of cases cultivation systems have a relatively small effect on the epidemic development of diseases and certainly much less than factors such as variety and weather.

Good mouldboard ploughing does not bury all the debris nor does straw-burning destroy all of it. Sufficient disease-bearing debris remains on the surface to initiate an epidemic if other factors are favourable. In only a limited number of cases (e.g. barley net blotch) is the additional debris left by non-mouldboard-ploughing systems sometimes a significant factor. However, the risk of serious slug damage to autumn-sown wheat is much greater in direct-drilled crops than in those following conventional ploughing.

There may be one exception to this general rule. The recently-observed disease barley yellow mosaic virus (page 205) is transmitted by a root-inhabiting fungus and early observations suggest that cultivation systems which drag and spread debris across the soil surface may result in a more rapid spread of the disease than does mouldboard ploughing.

Rotation

In cereals after grass, ley pests are important, but most root and stem diseases will be unimportant unless couch is common in the turf, when take-all may occur in small patches associated with the grass weed. If late-ploughed grass is quickly sown to autumn cereals, aphids may migrate directly from the grass to the seedling cereals. If they are carrying the virus, this will result in a severe attack of barley yellow dwarf. Cephalosporium leaf stripe has occasionally caused serious damage to cereals after ploughed-out grass and has sometimes been associated with inadequate control of wireworms.

Oats sown on lighter soils immediately after a run of wheat or barley are notoriously liable to damage from cereal cyst nematode while intensive or ‘continuous’ wheat/barley leads to few pest problems if stubbles are kept clean. Soil-borne diseases build up to a peak level over the early years of continuous wheat/barley and subsequently decline, while most leaf diseases seem not to be influenced greatly by rotation.

Wheat after fallow or roots is at risk to wheat bulb fly, to which oats are immune. Take-all is well controlled by a ‘break’ of roots or fallow unless stoloniferous grass weeds have intervened, while eyespot may be severe if the non-cereal ‘break’ is of only one season’s duration.

Sowing Date

Early autumn sowings are more prone to rusts, mildew, barley yellow dwarf virus, eyespot and take-all as well as being greatly at risk to such pests as *Opomyza florum*, frit fly and gout fly.

Sowing winter wheat later than mid-October increases the risk of wheat bulb fly damage but tends to decrease the incidence of disease. However, late sowing may incur serious yield penalties and a sensible compromise is invariably called for.

Late-sown spring barley is particularly susceptible to both mildew and brown rust, especially if it is adjacent to earlier-sown spring barley crops or to winter barley.

Spring oats sown after the end of March become progressively more vulnerable to frit fly damage, and every effort should be made to drill oat crops before this date.

Manuring

After the provision of optimum rooting conditions, adequate balanced manuring is the next essential for good cereal yields. Given adequate basal applications of phosphate and potash, the key to successful cereal production is optimum supply of nitrogen. An excess of nitrogen encourages greater development of some leaf and ear diseases, particularly mildew and rusts. A shortage may not only render the crop more susceptible to damage from some pests and soil-borne diseases, it also delays or prevents recovery from these attacks. Insufficient nitrogen is in itself damaging to yield, while over-generous supplies increase the risk of lodging and decrease grain size.

The use of fungicides and straw-shorteners enables crops to utilise nitrogen fertilisers more efficiently in producing higher yields, but they have only a small effect on the optimum rate of nitrogen application.

Parasites and Predators

Cereal pests and diseases have their own enemies which play an important part in regulating their numbers. Soil-inhabiting pests such as wireworms and leatherjackets are devoured by birds and some mammals. Slugs, aphids, wheat bulb fly and frit fly eggs are eaten by ground beetles, while the long white larvae of a fly called *Thereva* are predatory upon wireworms. Pests infesting cereal foliage also suffer from predation. Aphids (greenfly) are consumed in large numbers by hover fly grubs, adult and larval ladybirds, ground beetles and earwigs.

A glance at a wheat head infested with grain aphids will usually reveal at least one 'mummy', which is the remains of an aphid killed by a wasp-like parasite whose larva feeds internally on the aphid tissues. Other parasites of cereal pests include eelworms (nematodes), mites, fungi, bacteria and viruses. Some of the best-studied viruses of insects are those isolated from the body tissues of leatherjackets.

The reproductive potential of pests such as aphids is so great that under favourable conditions they outstrip the control exerted by parasites and predators and reach epidemic proportions, making chemical control economic. Nevertheless, it is worth encouraging beneficial organisms by using chemicals only when necessary, and then choosing ones which have the least harmful effects on parasites and predators (see ADAS Bulletin No. 20, *Beneficial Insects and Mites*). Recent experimental work has demonstrated the value of predators in checking the increase in numbers of cereal aphids in early spring.

Fungi, too, are known to have their parasites though the effect of these has not been much studied. An example is the occurrence of several different virus particles in the mycelium of the take-all fungus, though their presence has not been shown to affect the ability of the fungus to attack cereals in the United Kingdom.

Competition and Antagonism

Competition and antagonism between micro-organisms, especially in the soil, are recognised as important factors in the survival and activity of fungal pathogens. Attempts to harness them for improved disease control have not been very successful on a commercial scale though some control of take-all has been demonstrated in the laboratory and in small plots. Similar competitive effects may also occur with some cereal pests but their importance is not known.

Seed borne diseases

Standards of health and purity to type of cereals have been well maintained in Britain by seed merchants under the statutory UK Seed Certification Scheme controlled by the Ministry of Agriculture, Fisheries and Food. Seed purchased from merchants will usually be treated with disinfectant. For many years the standard treatment was organo-mercury which effectively controls diseases carried on the seed or in the seed coat. Loose smut, which is not controlled by organo-mercury, should not be a problem because of the legal standards set for this disease in seed which is sold. In recent years systemic fungicides have increasingly replaced organo-mercury as seed treatments. They are more expensive and need not be applied routinely, but they are effective against loose smut and (usually in combination with other chemicals) against all other seed-borne diseases. The low level of seed-borne diseases generally found in British cereal crops is evidence of the effectiveness of the measures applied in the seed schemes. Should they be abandoned to any extent there is a real danger that seed-borne diseases will again become important.

Many farmers have grown crops from home-saved seed and some make a practice of doing so for one or two years before purchasing new stocks from a seed supplier. Provided crops are carefully inspected for the important diseases and weeds and the necessary action is taken, no risks are entailed for one year, at least in the drier eastern parts of the country. Nevertheless, even for the first year, for autumn sown cereals particularly, and certainly if stocks are kept any longer, it is safer to arrange for proper cleaning and chemical treatment of the seed at least with an inexpensive organo-mercury treatment, as an annual insurance. Diseases which are not controlled by organo-mercury (loose smut of wheat, page 138, loose smut and possibly leaf stripe, of barley, pages 208 and 188) deserve particular attention when seed is saved on the farm.

CONTROL MEASURES: CULTURAL CONTROL

For all pests and diseases of cereals, modification of cultural treatments figures largely in control methods and, for many of them, this is the only method available. Many of the pests and diseases are sporadic and unpredictable in their appearance and therefore frequently the use of expensive routine chemical treatments is not cost effective. Furthermore, information relating pest or disease incidence to yield loss is often lacking, so preventing a useful economic appraisal of control methods.

Since many of the pests of cereals are classified as ley pests, which move on to cereals from grasses, much benefit derives from adherence to the rule of ploughing leys or old grass at least a month before the crop is due to be drilled. This allows time for drastic reduction in pest populations before the sown crop is available as an alternative food source.

Cereal stubbles infested with grass weeds are a potential source of trouble, inviting egg-laying by a whole host of cereal pests, and stubble hygiene should be vigorously practised immediately after harvest.

Similar considerations apply to many of the cereal diseases. Parasites, like mildew and rusts which can survive only on living host tissues, can be usefully restricted by stubble hygiene aimed at removing volunteers well before the autumn crop emerges. Volunteers can also act as a means of tiding pathogens such as *Rhynchosporium* and *Septoria* over the winter. Both these fungi and the net blotch pathogen also survive on debris from the previous crop and successful reduction of inoculum depends on efficient burial of debris, and this is often difficult to accomplish. Early surface cultivation, as soon as possible after harvest, is a main requisite for good hygiene. Before this, combine harvesters could be better adjusted to avoid high grain losses and the consequent large volunteer plant populations. Once shed grain has been induced to chit by mechanical cultivation, a judicious combination of cultivations and chemicals of the total weedkiller type, depending on the needs of the season, is likely to prove most successful rather than slavish adherence to any one method.

As mentioned previously, straw-burning, even a good all-over burn, will not reduce disease carry-over sufficiently to ensure a reduction in the severity of disease in the subsequent crop. Its main value is to enable the cultivator to move in early and without the hold-ups of straw-clogged tines. In addition, straw-burning can effectively remove a number of cereal pests; damage by frit fly larvae to early-sown winter cereals is often worse in those parts of the field where the preceding stubble had not been completely burned.

The effects of other cultural factors such as sowing date, manuring and rotations in modifying or controlling pest and disease outbreaks have been discussed earlier.

Resistant Varieties

With the intensification of cereals in Britain and the recognition of the importance of disease in limiting production, plant breeders include disease resistance as a major item in their programme in addition to their main aim of breeding for consistently high yield and quality.

One of the most valuable forms of resistance is that which allows disease escape, as in barley varieties which are resistant to loose smut because of the 'closed flowering' habit which prevents the fungal spores from reaching the flowers—the point at which infection takes place.

Most disease resistance, however, is not of this kind, but depends upon physiological factors. Such resistance is usually considered to be of two kinds, 'specific' (also associated with the terms race specific, vertical or major gene) and 'general' (also called incomplete or field and associated with the terms non-specific, horizontal or polygenic).

'Specific' resistance in plant breeding involves the use of one or a few genes which usually confer immunity or near-immunity to a particular parasite and provide a very high degree of control. This kind of resistance can be stable and long-lasting against some pests (e.g., cereal cyst nematode) and soil-borne pathogens which are relatively immobile. Experience has shown, however, that in the case of parasites such as rusts and powdery mildews which produce many spores and spread rapidly, causing epidemic disease, resistance of this kind is usually short-lived. This is because the pathogens are very variable in respect of pathogenicity factors and new varieties with new resistance of combinations of resistances select out appropriate virulent pathogenicity factors. In this way new strains (races) arise which are capable of overcoming the resistance of a new variety. Such strains build up rapidly as the new variety becomes widely grown and the variety becomes susceptible throughout the country within a short period, often within two or three years of being grown commercially. Sometimes when this happens, the variety turns out to be even more susceptible than the varieties it replaced.

With 'general' resistance plants become infected as do susceptible ones, particularly in the seedling state, but the rate of development of the disease is restricted as the plant ages (hence the term 'adult plant' resistance) so that disease levels and hence yield losses are much reduced. The level of disease actually achieved varies with environmental conditions but this kind of resistance has generally proved effective against all races of the pathogen and is usually considered to be more stable than is specific resistance.

'General' resistance, which manifests itself as field or incomplete resistance, has usually been associated with resistance derived from many genes (polygenic). However, the genetic basis for resistance is not always so clear-cut as that described above. There are, for example, cases where 'general' resistance is associated with one or a few genes and, on the other hand, specific resistance associated with a single major gene is not necessarily associated with resistance of a very high level.

Although many varieties, varying in the genetic basis of their resistance, have become susceptible to new races of pathogens, other varieties have been grown for many years under conditions favourable to diseases and yet have remained adequately resistant. For example, the winter wheat Cappelle Desprez was popular for many years and although occasionally, under severe disease pressure, it was severely attacked by yellow rust, under most conditions its resistance remained adequate and susceptibility to yellow rust was not a relevant factor when it was replaced by higher yielding varieties. Resistance of this type has been described as 'durable'. The description is based on the performance of a variety over a long period. It does not imply that the resistance is permanent or indicates how it manifests itself (as near-immunity or as 'general' resistance) or its genetic basis. Nevertheless, varieties that exhibit 'durable' resistance should provide a useful starting point for breeding for disease resistance which is likely to remain effective over a long period of time.

In the absence of many varieties with satisfactory and durable resistances to the major diseases, current varieties have to be used in a way which makes the best of the resistances available. To further this aim the concept of diversification of varieties has been introduced. For some of the major diseases, varieties are placed in diversification groups according to the genetic resistances they possess, and, therefore, the strains capable of attacking them. Crops to be sown in adjacent fields are selected from different diversification groups thus reducing the risk of rapid spread of disease from one field to the next.

This principle can be used to devise suitable variety mixtures to provide diversification within one crop. Experiments have shown that in suitable three-variety mixtures the severity of barley mildew can be reduced to about half of that expected when the varieties are grown separately and yield increases may approach those obtained from the control of mildew with fungicides. However, under severe disease pressure further yield increases would be expected from the use of fungicides on a mixture of varieties. Clearly an integration of the two approaches is desirable to obtain the most satisfactory results.

In the United Kingdom, diversification groups of varieties have been assembled for mildew of winter wheat and spring barley and yellow rust of winter wheat. Diversification tables are published annually by the UK Cereal Pathogen Virulence Survey and are available in such publications as *List of Recommended Varieties of Cereals* (National Institute of Agricultural Botany, see page 30).

While cereal diseases have rightly occupied the main attention of plant breeders, sources of resistance to pests have been utilised in cereal breeding programmes. Spring barley and spring oat varieties resistant to cereal cyst nematode and/or stem nematode are now available. Some winter and spring wheat varieties are known to be partially resistant to cereal aphids but these characteristics have so far proved to be difficult for incorporation into new varieties.

CONTROL MEASURES: CHEMICAL CONTROL

Chemical control is used as seed treatments for seed-borne diseases, barley mildew and seedling pests, as granule treatments for a few soil pests, as spray applications for severeal cereal leaf diseases or as emergency spray or bait treatments for pests such as aphids, leatherjackets and slugs.

Chemicals used for pest and disease control now must meet the requirements of the *Control of Pesticide Regulations 1986*. All reasonable precautions must be taken to protect human beings, plants and animals and to safeguard the environment, in particular to avoid polluting water. In addition a chemical must be used only for its 'approved use' as stated on the label or in a published list.

The conditions of approval relating to such matters as maximum dose, harvest interval, mixtures etc must also be observed.

Seed Treatment with Organo-mercury

For a long time organo-mercury seed disinfectants have played an essential role in controlling several serious seed-borne diseases. Although they have been replaced to some extent by systemic fungicides they remain an inexpensive means of controlling some important diseases. They are very poisonous, however, and some liquid preparations must be used only in premises registered under the Factory Acts. Other formulations are available to the farmer and should be used only as recommended. The instructions on the label should be followed implicitly. Unless reliable and efficient machinery is available the farmer is best advised to arrange for his seed to be treated by a seed merchant. Proprietary formulations are available which combine the organo-mercury disinfectant with an insecticide.

If the grain is damaged or damp when treated, subsequently stored in damp conditions, or treated at a higher than recommended rate, damage to the seed may ensue. This may result in failure to germinate or in the formation of abnormally swollen and stunted seedling shoot and roots, which fail to develop any further. If the seed treatment includes gamma-HCH the damage symptom is rather different, the shoot and roots showing a club or drumstick appearance, stunted and with the tips swollen.

The result of such damage is seen in the field as poor brairding. Good and bad drill widths occur, reflecting damage to particular sacks of grain or parts of seed lots.

Grain treated with organo-mercury and which is surplus to requirements may be suitable for sowing in the following season, provided it is of low moisture content and stored in good, dry conditions; the practice is not usually recommended. As an insurance, seed so kept should be submitted to an Official Seed Testing Station for germination tests before sowing. Treated grain must never be used for livestock feeding, even if diluted with clean grain. It should be well buried or burned or otherwise disposed of safely, away from access by livestock and wildlife.

In some countries organo-mercury treatments have been discontinued because of their poisonous nature. However, they have a long history as being cheap and effective and when used properly do not present a serious hazard. Alternatives are now available which, though more expensive, are much safer to use and are effective against a wider range of diseases. Furthermore the occurrence of resistance to organo-mercury in some pathogens (e.g. those causing oat leaf spot and barley leaf stripe) has made the use of other fungicides essential.

Other Seed Treatments

While organo-mercury compounds have been regarded as standard insurance treatments for cereal seed, other materials, particularly insecticides for the control of such pests as wheat bulb fly and wireworm, are applied to seed when necessary, either in combination with organo-mercury or separately. Insecticides used in this way include gamma-HCH, chlorfenvinphos and carbophenothion, to all of which reference is made later.

The traditional treatment for loose smut involving a laborious warm-water soak has now been superseded by the use of systemic fungicides which are usually used in combination with other fungicides to provide an effective control of all seed-borne diseases.

Some of the 'DMI' systemic fungicides (e.g. triadimenol, flutriafol) used against seed-borne diseases are also used to control mildew in barley and in fact were originally designed for this purpose. Carboxin, which was the first systemic to be used for the control of loose smut of wheat and barley remained effective for many years. However, recently the control of loose smut in winter barley has been unsatisfactory in some cases and this has been associated with resistance to carboxin.

Chemicals Applied as Sprays

A large range of chemicals is available for application as sprays to control pests and diseases in the growing crop. Most pests and diseases occur sporadically so that to attempt to control the whole range of pests and diseases by routine application of pesticides as prophylactics is not likely to be economically worthwhile. Provided care has been taken to select suitable varieties and to avoid high risk situations severe attacks of many pests and diseases can often be avoided. However, crops at high risk warrant prophylactic treatment and all crops should be monitored regularly and frequently so that outbreaks are identified at an early stage and the need for any application of chemicals can be assessed. For many pests and diseases criteria are being developed which take into account the likely development of the epidemic and the consequent benefit from the use of chemicals. Such criteria usually suggest the application of a treatment when a threshold stage is reached and other factors (e.g. weather, variety) tend to favour the development of an epidemic.

If prophylactic treatment is adopted then it should be planned in advance with a flexibility to allow for changes if outbreaks of particular pests or diseases occur. The plan should be devised to suit the need of each crop and the number of applications kept to a minimum for economic and environmental reasons. The cereal crop occupies a very large area and the use of chemicals should be confined as much as possible to those situations where they are necessary.

The Agricultural Development and Advisory Service (ADAS) have developed 'managed disease control' systems for each of the major cereal crops. These are based on information from experiments and field observation and provide guidance throughout the season on making decisions on fungicide treatments. The systems may include prophylactic treatments in some special cases but are mainly based on an assessment of disease risk in a particular crop (see ADAS publications, page 30).

Resistance to Chemicals

Sometimes the success of a treatment against a pest or disease may be partially or completely impaired because of the development of 'resistant' forms of the pest or fungal pathogen (in the case of the latter the terms 'tolerant' and 'insensitive' have also been used). Resistance may occur soon after chemicals have been introduced commercially or it may take many years of frequent use before resistance is detected. Usually the longer the chemical and the organism are in contact the greater is the risk of resistant forms emerging. Thus frequent use or long persistence of a chemical, or group of similarly acting chemicals, will increase the risk of resistance developing. The risk of resistance is greater still if the chemicals used act at a limited number of sites in the organism (site-specific) than if multi-site chemicals are used. A list of fungicides used on cereals, divided into site-specific and multi-site, and then grouped according to their mode of action is in a table on page 26. Forms which are resistant to one chemical are also resistant to other chemicals having the same mode of action (in the same numerical groups in the table). Thus in group 1, the 'MBC' fungicides, forms of the eyespot fungus resistant to benomyl are also resistant to carbendazim.

In the United Kingdom the resistance of some important cereal pathogens to some fungicides has become a problem in recent years.

Cereal fungicides: active ingredients grouped according to their mode of action†

Group number and name
Active ingredients

Site-specific fungicides

1. **'MBC' or benzimidazoles**
   benomyl
   carbendazim
   fuberidazole
   thiabendazole
   thiophanate-methyl

2. **'DMI' group of ergosterol biosynthesis inhibitors**

Imidazoles	imazalil prochloraz
Piperazines	triforine
Pyrimidines	nuarimol
Triazoles	flutriafol propiconazole triadimefon triadimenol

3. **Morpholine group of ergosterol biosynthesis inhibitors**

Morpholines	fenpropimorph tridemorph
Piperidines	fenpropidin

4. **Hydroxypyrimidines**
   ethirimol

5. **Carboxamides**
   benodanil
   carboxin

6. **Guanidines**
   guazatine

7. **Organophosphates**
   pyrazophos*

8. **Dicarboximides**
   iprodione

* More than one mode of action, exact classification uncertain.

Multi-site fungicides

9. **Phthalimides**
   captafol

10. **Phthalonitriles**
    chlorothalonil

11. **Dithiocarbamates**
    mancozeb
    maneb
    manganese/zinc dithiocarbamate complex
    'Polyram'
    propineb
    thiram
    zineb

12. **Mercurials**
    organomercury seed treatments

13. **Sulphur**
    sulphur

† From ADAS Booket 2257, *The Use of Fungicides and Insecticides on Cereals 1986*, Ministry of Agriculture, Fisheries and Food, Alnwick.

The first report of fungicide resistance was in the 1960s when strains of the oat leaf spot fungus were found to be resistant to organo-mercury. More recently there have been reports of resistance in the barley leaf stripe fungus to organo-mercury.

A more general problem has occurred in the powdery mildews. After the introduction of ethirimol (Group 4 in the table) in the early 1970s, there was a shift in the populations of the barley mildew fungus towards some resistance to the chemical. Whilst this resulted in some reduction in the effectiveness of ethirimol there was not a serious breakdown of disease control. Recent surveys after a period of reduced use of ethirimol suggest that there has been a shift back towards greater sensitivity to the chemical. During the 1980s there has been a build up of resistance in both the barley and wheat powdery mildew fungi to the 'DMI' fungicides (Group 2). Resistant strains are common and the effectiveness of the 'DMI' fungicides has been much reduced. Again, there has not been a complete breakdown of disease control but in some cases, especially in barley mildew, the level of control has been unsatisfactory. So far there have been no reports of resistance to the 'morpholine' group of fungicides (Group 3). In an attempt to overcome the resistance problem and to utilise the 'DMI' fungicides, which are also effective against other diseases, their use has been recommended in mixtures with fungicides with different modes of action, e.g., with fungicides in the 'morpholine group' (Group 3) or ethirimol (Group 4).

Strains of the eyespot fungus resistant to the 'MBC' fungicides (Group 1) were first noticed in 1982 and since then have become very common in crops of wheat and barley. In this case, the fungicides are completely ineffective in controlling the resistant strains and as a result another fungicide has had to be used for the control of eyespot in cereals. In 1985 resistance to the 'MBC' fungicides was found to be very common in *Septoria tritici* in England and Wales. Again, in this case, the 'MBC' fungicides give no control of disease caused by resistant strains.

Yet another case of resistance reported recently is that of strains of the fungus causing loose smut in winter barley which are resistant to the systemic seed treatment carboxin (Group 5).

With fungicide resistance becoming such a significant problem it has to be taken into account when fungicides are selected for the control of some cereal diseases. However, the more general problem of resistance has been recognised in both disease and pest control and farmers have been urged to adopt practices which are likely to minimise the chances of resistance problems occurring. Some of these practices are:

— Maintaining a low incidence and severity of pests and diseases by non-chemical means, e.g. by good husbandry, use of resistant varieties, use of diverse resistances.
— Avoiding routine prophylactic treatments except when they are essential.
— Where several applications are necessary to obtain control, using chemicals with different modes of action.
— In the case of fungicides, making full use of multisite fungicides (see table) which are less prone to fungicide-resistance problems and also making use of appropriate formulated fungicide mixtures or label recommended tank mixtures. To minimise the risk of resistance both fungicides in the mixture should be effective against the disease or diseases concerned and should be from different mode of action groups.

The failure to control a pest or disease with a chemical is not necessarily due to resistance. Such factors as the wrong choice of chemical, wrong timing, poor application and low dose rate should be considered before resistance is suspected.

Application of Chemicals

The extended use of fungicides and insecticides, coupled with the application of herbicides and fertilisers, has drawn attention to the need to rationalise application systems. One development has been the adoption of the 'tramline' system, so that all applicators use the same wheelways. These are made by blocking appropriate coulters on the seed drill, and provide access to the crop throughout the season. They make application much easier for the operator, minimise wheeling damage and result in more efficient and more accurate use of chemicals. The damage to crop yield from wheelings varies from negligible amounts in the early growth stages to about 3–4 per cent at the ear emergence stage. (This estimate is based on tramlines spaced for the use of a spray boom 12 m wide.)

Methods of applying pesticides as sprays have not changed for many years. The use of relatively large amounts of water (200–300 litres/ha) is time consuming and, together with the heavy equipment involved, limits the number of days available for spraying. Quicker methods of application involving less water and the use of controlled droplet size, or of electrostatically charged droplets are now being developed. Such methods will be adopted only if they are reliable and effective and also safe to the operator. Aerial applications, which use smaller volumes of liquid, are effective for the control of several pests and diseases, especially those affecting small plants or the upper parts of mature crops. However, where pests or diseases are well established at the base of a crop small volumes may be less effective.

Seed treatments are an effective and often the most economical way of applying chemicals; their use is necessarily prophylactic and may prove unnecessarily costly for controlling some pests and diseases.

PEST AND DISEASE INTERACTION

With so many pests and diseases to consider it is important to realise that, although they and their appropriate control measures must be described individually, many of them are likely to occur together on the same crop where they might need opposing cultural treatment for their control. For example later sowing may reduce the damage from eyespot, while at the same time the effects of wheat bulb fly will be made worse.

Some pests and diseases interact even more closely to the extent that one may be partially or even entirely dependent on the other. An example of the latter is cereal aphids as carriers (vectors) of barley yellow dwarf virus. Control of the virus disease can only be achieved by application of an aphicide to kill the aphid vectors (without which the virus could not spread at all). *Cephalosporium* leaf stripe of wheat seems incapable of entering the wheat plant to cause disease without the help of root-feeding pests such as wireworms. Take-all and cereal cyst nematode, on the other hand, compete with each other for root space and a heavy colonisation by either one seems to limit the development of the other.

With the expansion in the number of pests and diseases for which effective chemicals are available, it follows that their use may also be accompanied by unexpected effects on other pests and diseases. For example, some fungicides applied for the control of mildew have been accused of encouraging aphid multiplication by reason of their effects on fungi that parasitise them (though such effects have not yet been clearly demonstrated in commercial farming). Another example was provided by the 'MBC' fungicides, which, when used for eyespot control, tended to increase damage caused by sharp eyespot.

These examples serve to illustrate the complex nature of the pest and disease picture on cereals and the added complication of the various control methods available to the farmer. In practice a farmer must adopt a cropping system best suited to his own conditions. The system will include measures to control major pests and diseases, usually by cultural means though sometimes chemical treatments will have to be included.

The full effects of chemical control methods are sometimes slow to emerge and their use should be regarded as supplementing rather than replacing optimum soil and crop management which remain the important (and sometimes the only) means of ameliorating damage from pests and diseases. Such pests and diseases as occur are a natural part of the crop's environment and their containment within acceptable limits, not necessarily their elimination, should be the farmer's aim.

ADDITIONAL SOURCES OF INFORMATION

ADAS Leaflets on various cereal pests and diseases, HMSO and Ministry of Agriculture, Fisheries and Food Publications, Alnwick, Northumberland.

List of Recommended Varieties (Cereals), National Institute of Agricultural Botany, Huntingdon Road, Cambridge.

Barnes, H. F. (1956), 'Gall Midges of Economic Importance', Vol vii *Cereal Crops*, London: Crosby Lockwood & Son Ltd.

Colhoun J. (1971), 'Cereal Diseases', in *Diseases of Crop Plants*, ed J. H. Western, London: Chapman & Hall Ltd.

Empson, D. W. and Gair, R. (1982), 'Cereal Pests'. MAFF Reference Book 186. HMSO.

Jones, Gareth D. and Clifford, Brian C. (1978), *Cereal Diseases, their Pathology and Control*, BASF, Hadleigh, Ipswich.

Jones, F. G. W. and Jones, Margaret (1984), in *Pests of Field Crops* 3rd Ed., London: Edward Arnold (Publishers) Ltd.

Manners, J. G. (1971), 'Cereal Rusts and Smuts', in *Diseases of Crop Plants*, ed. J. H. Western, London: Macmillan.

McKay, R. (1957), *Cereal Diseases in Ireland*, Dublin: Sign of the Three Candles.

Moreton, B. D. (1969), 'Beneficial Insects and Mites'. MAFF Bulletin No. 20. HMSO.

Official Seed Testing Station, Agricultural Scientific Services, East Craigs, Corstorphine, Edinburgh 12.

Official Seed Testing Station, NIAB, Hungtingdon Road, Cambridge.

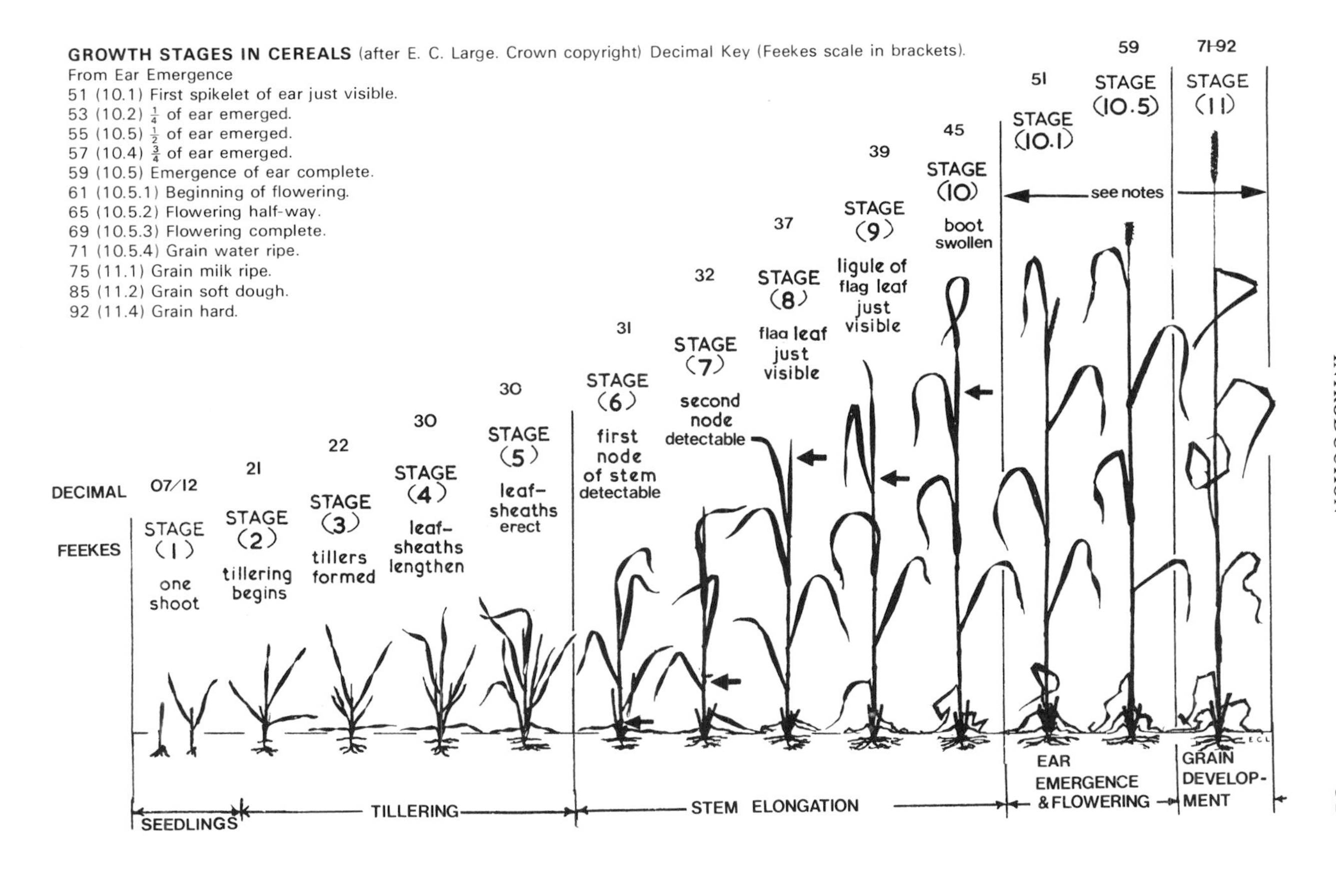
GROWTH STAGES IN CEREALS (after E. C. Large. Crown copyright) Decimal Key (Feekes scale in brackets).
From Ear Emergence
51 (10.1) First spikelet of ear just visible.
53 (10.2) ¼ of ear emerged.
55 (10.5) ½ of ear emerged.
57 (10.4) ¾ of ear emerged.
59 (10.5) Emergence of ear complete.
61 (10.5.1) Beginning of flowering.
65 (10.5.2) Flowering half-way.
69 (10.5.3) Flowering complete.
71 (10.5.4) Grain water ripe.
75 (11.1) Grain milk ripe.
85 (11.2) Grain soft dough.
92 (11.4) Grain hard.
DECIMAL
FEEKES
07/12
STAGE (1)
one shoot
21
STAGE (2)
tillering begins
22
STAGE (3)
tillers formed
30
STAGE (4)
leaf-sheaths lengthen
30
STAGE (5)
leaf-sheaths erect
31
STAGE (6)
first node of stem detectable
32
STAGE (7)
second node detectable
37
STAGE (8)
flaa leaf just visible
39
STAGE (9)
ligule of flag leaf just visible
45
STAGE (10)
boot swollen
51
STAGE (10.1)
59
STAGE (10.5)
71-92
STAGE (11)
see notes
ECL
SEEDLINGS
TILLERING
STEM ELONGATION
EAR EMERGENCE &FLOWERING
GRAIN DEVELOP-MENT

Chapter 2

WHEAT PESTS

PLANT DAMAGE SYMPTOMS

a. *Seed or seedling damaged at or before emergence*

Seeds missing from drill row	Birds (p. 75) Rats, mice (p. 77)
or hollowed out	Bibionid fly larvae (p. 48) Slugs (p. 68) Wireworms (p. 55) Mice (p. 77)

b. *Shoot develops but does not reach soil surface*

Shoot short and swollen, root tips club-shaped	Seed treatment injury (p. 23)
Shoot brown and shows signs of feeding....	Slugs (p. 68) Leatherjackets (p. 48) Wireworms (p. 55) Wheat bulb fly larvae (p. 36) Frit fly larvae (p. 44) Bibionid fly larvae (p. 48)
Shoot long and often twisted	Deep sowing (p. 84) Soil capping (p. 86)

c. *Seedling or tillering plant damaged*

i. Whole plants affected, usually yellow first

Plants pulled out and left on soil.........	Birds (p. 75) Rats, mice (p. 77)
Plants bitten near soil level	Bibionid fly larvae (p. 48) Slugs (p. 68) Leatherjackets (p. 48) Wireworms (p. 55)

	Grass moth caterpillars (p. 49) Swift moth caterpillars (p. 49) Cutworms (p. 49) Chafer grubs (p. 57) Wheat shoot beetle larvae (p. 58) Millepedes (p. 66)
Plants not bitten	Cereal cyst nematode (p. 74) Winter frost damage (p. 86) Fungus disease (p. 79) Virus disease (p. 134) Deep sowing (p. 84)
Shoot swollen or mis-shapen	Gout fly larvae (p. 44)
Plants slightly stunted, leaves spotted or flecked	Aphids (p. 60) Thrips (p. 65)
ii. Centre shoot yellows and dies (deadheart), outer leaves stay green	Wheat bulb fly larvae (p. 36) Frit fly larvae (p. 44) Late-wheat shoot fly larvae (p. 42) Grass moth caterpillars (p. 49) Common rustic moth caterpillars (p. 54) Grass and cereal fly larvae (p. 43) Yellow cereal fly larvae (p. 43) Bean seed fly larvae (p. 42) Wireworms (p. 55)
iii. As above, but a small neat hole often present near shoot base.................	Flea beetles (p. 59)

	Wheat shoot beetle larvae (p. 58)
iv. Leaves bitten, often well above ground	
Long narrow holes	Slugs (p. 68)
Roughly circular holes	Grass moth caterpillars (p. 49)
v. Leaves neatly cut off	
Horizontal bitten edge, tips missing..	Mammals (p. 77)
V-shaped leaf bite, tips often lying on soil..	Birds (p. 75)
vi. Leaves torn, with ragged ends	Slugs (p. 68) Leatherjackets (p. 48)
vii. Leaves shredded, reduced to fine fibrous mat	*Zabrus tenebrioides larvae* (p. 59)
d. *Plant damaged after tillering stage*	
i. Whole plant sickly, yellow or reddish	
Root system short, much branched, white cysts visible June onwards.......	Cereal cyst nematode (p. 74)
Root system with small thickened galls.	Cereal root-knot nematode (p. 74)
Root system normal.........................	Oat spiral mite (p. 66) Grass and cereal mite (p. 66)
ii. Some shoots yellow, others healthy......	Grass moth caterpillars (p. 49) Common rustic moth caterpillars (p. 54)
iii. Some shoots swollen, others healthy	Gout fly larvae (p. 44)
iv. Stem surface near upper nodes pitted with saddle-shaped depressions	Saddle gall midge larvae (p. 47)
v. Leaves bitten or discoloured	
Long narrow strips eaten from leaf blade	Slugs (p. 68)

Symptom	Pest
	Cereal leaf beetle (p. 58)
	Barley flea beetle (p. 164)
Irregular areas bitten from leaf edge....	Grass moth caterpillars (p. 49)
	Leaf sawfly larvae (p. 59)
Discoloured blotches on leaf surface	Aphids (p. 60)
Silver marks on leaf surface................	Thrips (p. 65)
Leaves with blister mines..................	Cereal leaf miner larvae (p. 45)
vi. Shoots bend or break and fall over	
Shoot bent above ground level, brown seedlike bodies beneath leaf sheath near cut edge................................	Hessian fly larvae (p. 161)
Shoot cut off cleanly just above ground level..	Wheat stem sawfly (p. 59)
Shoot cut off with diagonal bite or peck mark, heads often stripped.............	Birds (p. 75)
	Mammals (p. 77)
Shoot breaks off near upper nodes	Saddle gall midge larvae (p. 47)
e. *Damage to flowering heads*	
Tips of ears white, distorted or shrivelled...	Frit fly larvae (p. 44)
Whole ear white or aborted.....................	Grass and cereal mite (p. 66)
	Hessian fly larvae (p. 161)
	Wheat stem sawfly larvae (p. 59)
Ears distorted, stems often twisted	Wheat gall nematode (p. 71)
	Oat spiral mite (p. 66)
	Frit fly larvae (p. 44)
Ears often one-sided, groove runs down side of stem......................................	Gout fly larvae (p. 44)
Ears sticky, often covered with extruded fluid...-	Aphids (p. 60)

f. *Damage to ripening grains*

Grains missing......................................	Lemon wheat blossom midge (p. 46) Birds (p. 75)
Grain shrivelled or incompletely developed.	Grain aphid (p. 62) Orange wheat blossom midge (p. 46) Thrips (p. 65) Birds (p. 75)
Grain blackened and/or partly eaten	Rustic shoulder knot moth caterpillars (p. 54)
Grain replaced by black powdery mass or purple-black galls...............................	Wheat gall nematode (p. 71) Frit fly larvae (p. 44)
Grain flattened, white deposit on outside ...	Birds (p. 75)

INSECTS

Wheat bulb fly (*Delia coarctata*) (Plates 1–5)
This is the most common insect pest of winter wheat and is habitually present in localised areas in the eastern half of England and Scotland. Attacks range westward as far as Warwickshire, Gloucestershire and Wiltshire, although the adult flies have been found even further west. Winter cereals in many central and north European countries, including Russia, suffer from the attention of this pest.

Damage symptoms
Occasionally wheat shoots are destroyed before reaching the soil surface; this is often the case when spring wheat is drilled early (i.e. before mid-March) on infested fields. Late-sown or backward winter wheat plants may suffer similar damage, the shoot turning brown as a result of feeding by the young maggots.

More usually, damage occurs to winter wheat from the seedling stage to the end of tillering. The centre leaves turn bluish then yellow and die ('deadheart' condition), while the outer leaves remain green. Slitting open such affected tillers in the period from late February to early May will usually reveal the white, legless maggot feeding near the stem base.

Life history
Adult flies resemble house flies but are rather smaller. They are on the wing from mid-June until September, the males emerging from five days before the females. Both sexes remain on the wheat crop from which they emerge for about three weeks, then the female flies move to other fields and hedgerows nearby.

Creamy white eggs are laid from mid-July until early September, the peak of egg-laying usually occurring in late July on clay soils and a fortnight later on peat soils. Each female fly lays about 50 eggs in bare soil, usually in crevices and just below the soil surface. Land at risk to egg-laying includes full fallows, bastard fallows, early potatoes, vining peas, sugar beet and other root crops. A rough tilth favours egg-laying, and more eggs are laid in dry than in moist soil.

The eggs do not hatch until mid-January at the earliest, so that they remain in the soil for a period of approximately six months. Hatching may be delayed until March by spells of very cold weather, and in northern counties late February is the usual time when hatching commences.

The white, legless larvae which emerge from the eggs are attracted by chemical secretions to plants of wheat, barley, rye and a number of grasses including couch. Each maggot bores into the centre of a shoot just below ground level. The stem tissues turn brown and the central leaf yellows and dies. Towards late March or early April the maggot, now much larger, moves into a new shoot on the same or a neighbouring plant, producing the same 'deadheart' symptoms. Farmers usually notice attacks at this stage, when patches occur in the crop which takes on a bluish, unthrifty appearance.

The maggots continue to feed until early May when they leave the plants for the surrounding soil. There the larval skin hardens to form a brown puparium, easily noticed in dark coloured soils. Adult flies emerge from these puparia in mid-June in most seasons, so that there is only one generation each year.

Effect on yield
In years when egg and larval numbers are high, whole fields of winter wheat can be devastated. Similarly spring wheat sown too early on infested land can be completely destroyed.

More usually, attacks lead to severe thinning of the crop in March or April, and the effect on crop yield depends on the ability of the plants to recover in late spring and early summer. Obviously late-sown, backward crops which have hardly begun to tiller when

Plate 1. Wheat bulb fly. Adult female (×7).
A Shell photograph.

Plate 2. Wheat bulb fly. Adult male (×7).
A Shell photograph.

Plate 3. Wheat bulb fly. Eggs (×10). A Shell photograph

Plate 4. Wheat bulb fly. Larval damage to winter wheat. Seedling on left attacked before or about the time of emergence. Plant on right attacked after emergence. Crown copyright.

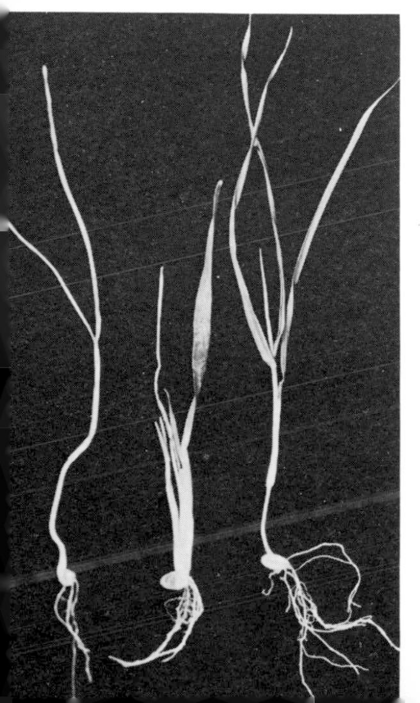

Plate 5. Wheat bulb fly. Larva (×8) dissected from tissues of winter wheat shoot. Crown copyright.

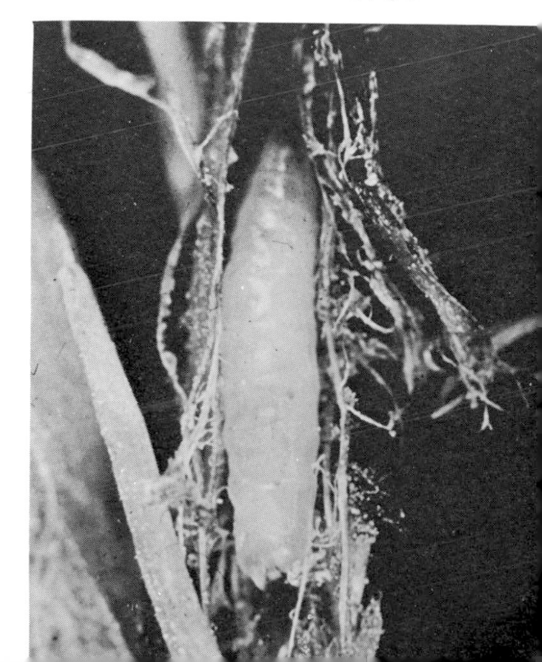

the attack starts are most susceptible to damage and are least able to compensate.

While the areas in which the pest habitually causes trouble are often localised, it is believed to be responsible for the loss of 16,000 equivalent hectares on average each year in England alone. This figure fluctuates widely between years depending on egg and larval numbers and the state of crop growth. In 1953 direct losses from attacks in an epidemic were estimated at £1.25 million (about £5 million at current values).

Control

Each stage in the life-cycle is subject to natural mortality factors which considerably reduce numbers surviving to adulthood. Thus between 50 and 80 per cent of the eggs laid are infertile, dry up or are eaten by beetles and other predators. The larva can survive in the soil for only a short period and dies if it cannot quickly find a host plant. In all, about 70 per cent of larvae die and further deaths occur when insufficient shoots are available for colonisation. Puparia may be destroyed by predators and parasites, while adult flies are attacked by spiders, dung flies and fungi. It seems that in most years only about 40 per cent of the eggs laid survive to produce larvae within shoots.

Cultural control measures are of primary importance and aim at producing strong, well-tillered plants by early spring. Early drilling of winter wheat, not later than mid-October, on a firm seedbed adequately provided with phosphatic fertiliser gives plants which can afford to lose one or two tillers to bulb fly attack but which still produce enough fertile shoots. By contrast, late-sown or backward crops contain plants which have either failed to tiller or have produced few tillers by February, and these plants may be killed outright. Rolling and top-dressing with nitrogenous fertiliser help attacked crops to recover, and should be applied as soon as possible after the onset of damage.

Of course, trouble can be avoided by not growing winter wheat in fields known to be at risk. Such fields can be identified by pre-cropping egg counts. Advice is available on the general likelihood of attacks in any area for each season. Oats are immune to damage and can be grown instead of wheat. Winter barley, although initially attacked, does not seem to be a favoured host and few larvae survive within barley shoots. However, attacks by larvae to the seedlings below ground during establishment can result in patchy crops. Spring wheat drilled from mid-March onwards is not affected by the pest, and provided other factors such as disease risk, spring labour demands and economics are not limiting, a switch from winter- to spring-sown wheat can provide a reliable means of avoiding wheat bulb fly damage.

Working fallow land to a fine tilth which is then left stale during the egg-laying period might reduce the numbers of eggs laid in the soil. Deep ploughing may bury some eggs, and increase their mortality. Sowing a catch crop of mustard or rape after early potatoes or early vining peas might provide a cover for the land during the egg-laying period. However, germination failures are so frequent during June that the methods cannot be recommended with confidence.

Chemical measures give only partial control but usefully supplement the cultural measures outlined above. Seed treatments with insecticides using higher rates than those recommended for wireworm control are needed. Most seed treatments are more effective when the seed is drilled shallowly.

Carbophenothion, chlorfenvinphos and fonofos are currently available as seed treatments, usually combined was one of the normal organo-mercury fungicides. They prevent many grubs from entering the plant, and in addition they kill some of those which succeed in penetrating the shoots.

Cereal seed treated with any of these insecticides should be of high quality and with a moisture content below 16 per cent. Treated grain should be kept in a dry store and for no longer than necessary.

Treated seed constitutes a risk to grain-eating birds, especially during early spring. For this reason chlorfenvinphos and carbophenothion may be used only on wheat sown before the end of December. Carbophenothion must not be used in Scotland. When drilling, care should be taken to avoid spillages and too shallow drilling. As a general rule, insecticidal seed treatments for wheat bulb fly control should be restricted to those areas where there is a definite risk from damage by the pest.

As an alternative to insecticidal seed treatment, particularly in northern counties of England where seed treatments are often ineffective and phytotoxic, fonofos granules can be applied to the soil surface immediately before drilling. Similarly, sprays of chlorfenvinphos, chlorpyrifos or fonofos can be applied at, or immediately after, drilling. Fonofos is also effective against wire-worms. At the time of egg hatch (mid-January to March), a spray application of chlorpyrifos, fonofos, pirimiphos-methyl or triazophos can give useful results.

For unprotected crops which show symptoms of attack from late February onwards, a single insecticidal spray of dimethoate or omethoate can kill many grubs within the shoots and give an economic yield response. Sprays applied before mid-March are usually more effective than later ones. Farmers are strongly advised to examine from late February onwards those wheat crops which may be at risk. However, dissection of shoots in the laboratory gives the most reliable indications of the need to spray.

Crops which are likely to be subjected to heavy larval damage often respond to a combination of two or more of the chemical control methods outlined above. Such multiple treatments have been shown to give a worthwhile economic return.

Late-wheat shoot fly (*Phorbia securis*)

The damage caused by this insect closely resembles that caused by wheat bulb fly (see page 36). Attacked plants show the typical deadheart condition in which the central leaf of an infested shoot turns yellow and dies while the outer leaves remain green. Damage in winter wheat occasionally overlaps that done by wheat bulb fly but the white maggots of *P. securis* are very small in late April or early May when those of wheat bulb fly are fully grown.

The life history usually involves one generation a year, although a second generation has been found in southern England on Yorkshire fog grass. Adult flies emerge in March and April and lay their eggs near the edges of the leaf sheaths of winter wheat, spring wheat and barley. The white, legless grub enters the centre of a shoot and feeds for about a month without migrating to other shoots or plants. The full-grown grub pupates in the soil.

Damage is not considered to be of economic importance and no control measures are required against this fly.

Bean seed flies

A complex of species, including *Delia platura* and *D. florilega*, occasionally feed on winter cereals as well as on a wide range of other plants.

The white, legless maggots feed on germinating cereal seedlings or nibble the shoots and leaves of older plants. There are usually three or four generations each year but the autumn generation is the only one affecting cereals in this country. The adult flies are attracted to freshly ploughed or heavily manured land and attacks on cereals usually follow a ploughed crop of rape or mustard.

Bean seed flies are more important on runner beans, marrows and other vegetables than on cereals in this country. This contrasts with the situation in the USA, where seed-corn maggots (as they are known) are economically damaging to cereals.

Grass and cereal flies

A number of different flies such as *Cetema*, *Geomyza*, *Opomyza*, *Meromyza* and *Oscinella* are grouped together under this heading, for they share certain common features. The larvae are white, yellowish or green and spend the winter inside grass shoots. They are able to migrate into cereals if infested grass is ploughed shortly before drilling a cereal crop. Attacked grasses and cereals show the 'deadheart' condition typically produced by stem-boring pests. Because damage to cereals frequently follows the ploughing of grassland (ley or permanent grass) these flies are often collectively known with grass moths, leatherjackets, frit fly, wheat shoot beetle and wheat flea beetle (q.v.) as ley pests.

Cultural control measures rely on ploughing grass at least a month before sowing winter cereals, and by ploughing grass in the autumn preceding a spring cereal crop. Cereal stubbles should be kept free of grass weeds for these attract the egg-laying female flies.

Chemical control measures have been developed for some species. Identification of the maggots requires expert knowledge.

Yellow cereal fly (*Opomyza florum*)

This fly sometimes causes trouble in early-sown winter wheat and to a lesser extent in winter barley and triticale. Crops sown before mid-October are at greatest risk, and since 1979 attacks on winter wheat have been particularly noticeable in East Anglia and south-east England.

Damage symptoms are superficially similar to those caused by wheat bulb fly and in fact maggots of the two species may be found together in the same crop. *O. florum* larvae are rather smaller and pointed at both ends. They climb up the plant and move downwards within the leaf sheath to make a brown circular or spiral feeding mark at the base of the shoot or just above the first node. The larva does not migrate from infested to healthy tillers and pupates within the infested shoots. In contrast, wheat bulb fly larvae are bluntly truncated at the hind end; they bite a large, ragged hole in the side of the shoot, often move from infested to healthy tillers and even to neighbouring plants, and pupate in the soil.

O. florum adults are on the wing from June onwards, moving from infested crops into hedgerows and woodlands where they become sexually mature by October. The gravid female flies move into brairding cereal crops to lay their eggs on the soil close to the plants. The eggs hatch in late January, February or March in a process governed by prevailing soil temperatures.

The status of *O. florum* as a pest of cereals is still uncertain. In most years the number of tillers damaged by larval feeding is insignificant and may even be unnoticed by the farmer. However, even when a large percentage of tillers are destroyed (as in 1979–81 in some crops in East Anglia), the wheat crop seems able to recover and yield satisfactorily. Comparisons of attacked and protected crops show that the plant responds to *O. florum* damage by producing further tillers, many of which are fertile but which bear small ears containing fewer grains per ear and smaller individual grains than those in unattacked plants. This explains why chemical control measures, optimally applied as single spray treatments at the start of egg hatch, effectively protect the crop from tiller damage but do not always give an economic yield response. Insecticides which have manufacturers' recommendations include chlorfenvinphos, fonofos, omethoate, triazophos and several synthetic pyrethroids.

More work is needed to establish thresholds of egg and larval numbers above which yield is likely to be adversely affected and thus to determine which crops need chemical control treatment.

Gout fly (*Chlorops pumilionis*) (Plates 49–51, pages 158–9)
Attacks by gout fly are for reasons unknown a good deal less frequent nowadays. Wheat, barley and rye are its cereal hosts, oats and maize being immune.

Damage symptoms in wheat closely resemble those in barley and rye, and are described together with the life history under barley pests (page 157).

Control measures for autumn-sown wheat depend on drilling not later than October in a seedbed adequately supplied with phosphatic fertiliser. Spring wheat should be drilled before the end of March to escape attack.

Frit fly (*Oscinella frit*)
This insect will be considered in detail under oat pests (page 216).

Attacks on wheat usually occur in autumn or winter when wheat follows grass or grass-infested cereal stubbles in the rotation. Damage is much worse in early-sown crops and in those parts of the field where the preceding stubbles were incompletely burned. Attacked shoots show the usual 'deadheart' symptoms from December onwards with a yellow centre shoot, containing the thin pointed maggot, surrounded by green outer leaves.

The amount of damage depends on several factors including the numbers of frit fly larvae in the grass at ploughing, the interval between ploughing and drilling, the length of time soil temperatures remain above the threshold of activity of the larvae in autumn (7·2°C), and whether the grass is cut for hay or silage or is grazed (grazed swards generally having greater numbers of larvae).

Recent changes in husbandry such as direct drilling, minimum cultivations and shorter intervals between ploughing and drilling have increased the likelihood of frit fly damage.

A later generation may affect the flowering heads of spring wheat, the florets at the tip of the ear appearing white and empty. As similar damage can be caused by late frosts, the presence of frit fly grubs or puparia in affected ears is needed to confirm that the pest is responsible.

Attacks on winter wheat are avoided when preceding grass is ploughed in the second half of August and this also circumvents wheat bulb fly damage. Frit fly eggs are often laid on volunteer cereal plants or on grass weeds, and stubbles should be ploughed early and kept as clean as possible using cultivations or suitable herbicides. There is some evidence to suggest that frit fly females occasionally lay eggs direct on seedlings of early-sown winter wheat.

Winter wheat crops at risk may be treated with a spray application of chlorpyrifos, fonofos, omethoate or triazophos in late autumn. Treatment with insecticides after 'deadheart' damage has been seen is of little value.

Leaf miners (*Agromyza* spp., *Pseudonapomyza* sp., *Cerodontha* sp.)

The flag and upper leaves of winter and spring wheat and barley may show blotchy mines, caused by maggots of at least six kinds of leaf mining flies. The grubs bore into the central tissue of the leaf, leaving both upper and lower leaf surfaces intact, so that a pale blister is formed. Tissue extending outwards from the blister to the leaf tip turns yellow and dies. After becoming full-grown the larvae of some species bore out of the mine and fall to the soil. Feeding punctures made by adult flies are sometimes noted on seedlings during the autumn. They are visible to the naked eye as white dots, often arranged in rows along the leaf.

Little is known of the economic consequences of leaf miner damage, which was prevalent in the mid-1960s. All cultivars of winter and spring wheat seem susceptible.

Wheat blossom midges

Widespread attacks by wheat blossom midges are rare and the interval between attacks in this country is often twelve or more years. Several fields in northern counties of England have been sprayed in the past few years to control these pests.

Damage symptoms

The developing grains within the flowering heads of winter or spring wheat either fail to develop or are shrivelled in appearance. Small lemon or orange maggots can be found within the ears for about a month after flowering.

Life history

Two distinct species are involved in attacks on wheat. Grubs of the lemon wheat blossom midge (*Contarinia tritici*) are lemon-yellow and often occur in numbers up to ten in each attacked floret. They feed on the young ovule so that the floret becomes completely blind. Adult midges emerge from the soil in the mid-summer, mate and quickly lay eggs in small groups on the flowering wheat ear. The flight period occupies about four weeks and the midges fly in calm, dry weather conditions. The eggs soon hatch and the maggots feed for about three weeks before jumping from the ears to the soil. There each larva spins a cocoon in which it spends at least one, sometimes up to four, winters. Those maggots which leave their cocoons in spring migrate upwards close to the soil surface, pupate for three weeks and emerge as adult midges.

Maggots of the orange wheat blossom midge (*Sitodiplosis mosellana*) are orange or reddish in colour and are usually found singly in each attacked floret. Their feeding causes the grain to be shrivelled and of poor quality. The life history differs from that of *C. tritici* in that the larvae remained within infested ears. Larvae may remain in their cocoons in the soil for at least ten years; some for as long as nineteen years.

Control

DDT was used in Eire to control midge attacks in an epidemic which occurred in 1951 but is no longer permitted in the United Kingdom. The timing of an insecticidal application is critical and has to be made immediately before ear emergence. Damage occurs so infrequently that the routine use of an insecticide cannot be justified. In parts of the Yorkshire Wolds attacks have increased in severity in recent years and spray applications of fenitrothion or chlorpyrifos have given worthwhile yield responses.

There is some evidence of varietal resistance to midge damage, but this has not been exploited by plant breeders. Varieties of wheat which are particularly susceptible to attack and which should not be grown in high-risk areas include Hobbit, Hustler, Sicco and Sportsman (orange midge) and Aquila, Chalk, Kador and Mega (lemon midge).

Saddle gall midge (*Haplodiplosis marginata*) (Plate 6, page 50)
This insect is an important pest of cereals in northern continental Europe. Although it has been recorded frequently in the British Isles since 1889, only in recent years has its occurrence given cause for some concern. Wheat and barley grown in close cereal rotations or heavy clay soils in localised areas of East Anglia and the east Midlands are chiefly at risk, but even here number of over-wintering larvae have dropped sharply since 1970.

Host plants include wheat, oats, rye and a number of grasses including couch grass.

Damage symptoms
Blood-red legless maggots feed in depressions on the stem beneath the leaf sheaths of the first or second internodes. The galls can often be felt through the upper leaf sheath as knobbly outgrowths. When attacks are severe the stem may be so weakened at the saddle-shaped galls that it breaks and the ear falls to the ground.

Life history
Adult midges emerge from the soil from late May to early June, chiefly in periods of warm, showery weather. Eggs are laid in batches on the leaf surface and form red lines which can be seen with the naked eye. The young larvae migrate into the leaf sheaths and feed on the stem tissue for a few weeks before entering the soil. There the grubs either remain as free larvae for one or more winters or spin cocoons from which they emerge in late spring, pupate for a few days and change to adult midges.

Control
As the midges do not move far from the field in which they emerge, a simple rotational method of control is to introduce one or two break crops in place of wheat or barley. Insecticidal sprays of DDT, parathion or (more recently) fenitrothion are commonly used on the European Continent and give useful yield increases

when attacks are severe. However, only the latter is permitted in the United Kingdom. The spray has to be applied about ten days after the peak flight period of the adult midge, i.e. in the first half of June.

Assessment of damage and of the likely benefits of insecticidal treatment under British conditions have been investigated. Critical levels of infestation for wheat are about twelve million overwintering larvae per hectare in soil, or twenty-two larvae per attacked stem. Apart from the loss of ears larval feeding reduces the size of grain in attacked shoots.

Hessian fly (see page 161)

Bibionid flies

There are several British species, the most common being the St Mark's fly, *Bibio marci* (L.), the horticultural fly, *B. hortulanus* (L.) and the fever fly, *Dilophus febrilis* (L.). *Bibio johannis* (L.) has also been reported as damaging young cereal plants. These are robust flies, often pubescent with shorter legs and wings than any other members of the Nematocera. Some species exhibit sexual dimoprhism with reddish-brown females. The St Mark's fly is a large black hairy insect (about 10 mm long), often to be seen in large numbers around the edges of grass fields, especially during the first warm days of May. The legs hang down below the body during flight. The horticultural fly is a smaller black insect (5–7 mm long) which emerges from April to June.

Eggs are laid in clusters just below the soil surface in grassland or in soil rich in organic matter. The maggots remain clustered together in the soil. They have distinct brownish or black heads with biting jaws, and reach a length of up to 25 mm depending on species.

The larvae are often found in grassland nibbling at the grass roots, but on occasions they cause severe damage to cereals, especialy winter wheat or barley. The older larvae hollow out seeds in a manner reminiscent of slug damage, or they sever the lateral roots and/or stems just below ground.

Chemical control measures have not been developed for use against bibionid fly larvae.

Leatherjackets

These 'ley pests' will be considered under barley pests (page 162). The symptoms of damage to both winter and spring wheat and the control measures to be adopted are the same as for barley.

Swift moths (Plate 7)
Caterpillars of the ghost swift moth (*Hepialus humuli*) and the garden swift moth (*H. lupulina*) feed on wheat plants about 2·5 cm below soil level, cutting through the stems of a number of adjacent plants along the row.

The larvae are 2·5 cm or more in length when full-grown, white with brown shiny heads. Those of the garden swift moth feed during the autumn and winter, and the life cycle takes only one year. Caterpillars of the ghost swift moths attack cereals less frequently and take two years to complete their life cycle. Adult moths of both species are dingy brown and fly in summer. They lay eggs in grass or in weed-infested crops.

Removal of grass weeds from land about to grow cereals is a useful cultural method of control. Chemical measures are rarely adopted because by the time the damage is usually seen it is too late for insecticides to help.

Cutworms (Plates 8–11)
In hot, dry summers a number of noctuid moths may fly in large numbers. Their caterpillars called cutworms are characteristically thick with dark markings along a brown or green body and they often lie curled up at or just below the soil surface.

Cutworm attacks are more important on vegetable crops such as celery, lettuce, red beet, carrots and potatoes. Wheat crops are damaged infrequently, several plants in the row being cut off below ground level. The foliage turns yellow and the cutworm responsible can usually be found close at hand.

Control measures are not usually needed for wheat.

Grass moth (Plate 12)
Several species of grass moth live in this country, and they can often be seen fluttering weakly over grassland at dusk in late summer. Only one species is known to damage cereals, the crambid moth (*Agriphila straminella*). This is quite different from the North American scene, where several grass moth species are of economic importance.

Attacks by the crambid moth larvae usually occur in winter or spring wheat following permanent or temporary grass or a herbage seed crop. In late spring the caterpillars eat small circular holes in both stems and leaves. Later whole shoots may be bitten off at ground level, the yellowing leaves lying strewn on the soil surface. Older larvae may enter wheat shoots and produce ‘dead-

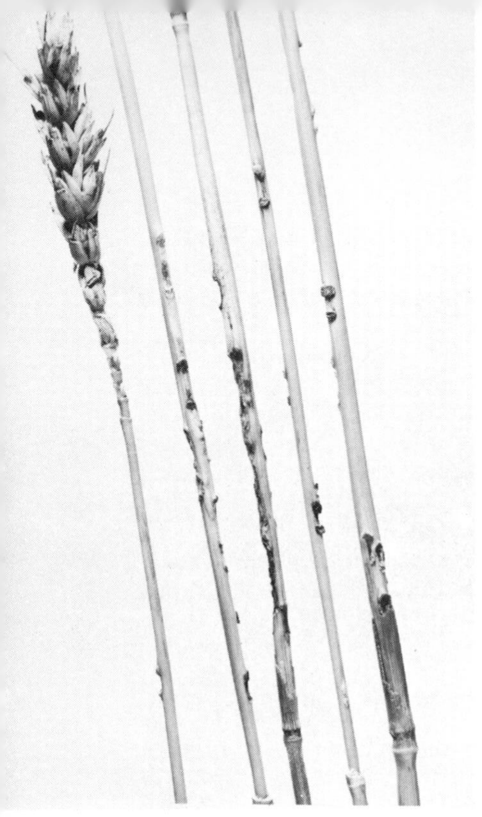

Plate 6. Saddle gall midge. Larval damage to wheat stems, showing saddle-shaped galls.
National Institute of Agricultural Botany.

Plate 7. Swift moth. Full grown caterpillar (×2). Crown copyright.

Plate 8. Large yellow underwing moth (×2½) whose larva is one of the most common cutworms found in Britain. Crown copyright.

Plate 9. Large yellow underwing moth. Eggs (×18) on debris. Crown copyright.

Plate 10. Large yellow underwing moth. Pupa (×6).

Plate 11. Large yellow underwing moth, larva or cutworm (×5) curled in characteristic position on soil surface.
Crown copyright.

Plate 12. Grass moth (Crambid). Full grown caterpillar with damaged wheat plant in background.
Crown copyright.

heart' symptoms reminiscent of common rustic moth caterpillars or other stem borers.

Careful search in the soil below attacked plants may reveal the purplish, spotted caterpillars, about 1·3 cm long, lying within silken tunnels attached to the base of the plants. The tunnels are often lined with green faecal pellets.

The caterpillars feed on grass or cereals during autumn and spring. In May they spin cocoons within the tunnels, pupate and emerge in June or July as small whitish moths.

Early ploughing of infested grassland helps to minimise serious damage by this 'ley pest'.

Common rustic moth (*Mesapamea secalis*)

Another 'ley pest' which frequently attacks winter wheat following grass or a grass-infested stubble is the caterpillar of the common rustic moth.

Groups of plants with yellowing central leaves can be seen in March and April. Careful inspection of these shoots will often reveal the caterpillar feeding within the lower portion. Even if the larva has already moved out of the shoots, its characteristic dark green faecal pellets remain to identify it. The caterpillar grows to about 1·8 cm long and is green with two brownish stripes running along the back of the body. Pupation occurs in the soil. The brownish adult moths fly from June until September and lay their eggs mostly on grasses but also occasionally on autumn-sown wheat and other cereals.

Cultural control depends on allowing a margin of at least four weeks between ploughing grass and drilling a following cereal crop. Rolling and applying a nitrogeneous top-dressing in spring helps attacked crops to recover.

While the pest may give rise to severely thinned patches in winter wheat fields, its attacks are seldom if ever sufficiently serious to warrant re-drilling of the crop. Chemical control measures have therefore not been developed.

Other moth caterpillars

Caterpillars of a number of moth species feed on wheat plants. Many are of little or no economic importance, and include the rustic shoulder knot moth (*Apamea sordens*), the marbled minor moth (*Oligia strigilis*), the rosy rustic moth (*Hydraecia micacea*), the flounced rustic moth (*Luperina testacea*), the Brighton wainscot moth (*Oria musculosa*), and a tortricid moth (*Cnephasia longana*). At least one of them (*O. musculosa*) is of some importance in cereal growing areas of France and Russia.

BEETLES

Relatively few kinds of beetles attack wheat in the United Kingdom. They do so either in the immature larval stage or as adult beetles or as both, for the larvae as well as adults possess strongly developed biting mouthparts.

Wireworms (Plates 13 and 14)
These are the well-known larvae of click beetles; the term 'click' describes the way in which the beetles, when placed on their backs, jump into the air by bending and straightening the body.

Damage symptoms
Wireworms bite through young wheat plants below ground level, so that the whole plant turns yellow and dies. Larger plants may have occasional leaves yellowed, but in both cases the damaged plant tissues clearly show the ragged edges associated with the tearing action of the larval mouthparts. Several plants along a drill length may be damaged, with the wireworm responsible lying near to the next healthy plant. Large patches of attacked plants may appear, usually towards the centre of the field.

Life history
Click beetles belong to a number of species, of which *Agriotes lineatus*, *A. obscurus*, *A. sputator*, *Athous haemorhoidalis* and *Ctenicera* spp. are the most common. These are all dull brown insects measuring 0·6 to 2·5 cm in length, and grassland is their natural habitat.

Adult beetles feed on grasses and cereals between April and July but do little harm. Eggs are laid in midsummer in small groups in the soil. These hatch after a few weeks and the young wireworms, almost white, commence feeding. Larval development takes four or five years to complete, during which time the larva turns to a bright yellowish-brown. The long cylindrical body with its dark brown head and three pairs of short legs can be clearly seen.

Larval feeding reaches peaks of activity in late spring and again in early autumn, probably because the wireworms move deeper into the soil in midsummer and in winter. When mature, the larva pupates in a hollow cell in the soil in July or August. The adult

Plate 13. Adult click beetle (×10) whose larva is one of our most common wireworms.

Plate 14. Wireworms (×3) of different ages all of bright orange colour.

beetle emerges from the pupal case within a few weeks, but usually remains in the cell until the following spring.

Control

Until a few years ago, wireworms were rightly regarded as a major threat when old grass or a long ley was ploughed for cereals or other arable crops. In recent years wireworm numbers have declined throughout the country, even in the heavy clay soils so favoured by them; it is now extremely uncommon to find much wireworm damage to individual plants, while severe damage to whole crops is virtually unknown in England and Wales. The reasons for this decline are not fully understood, but may be associated with the widespread use of persistent soil insecticides.

Much of the seed wheat drilled every year is treated with an insecticidal seed dressing as an insurance against damage from infestations ranging up to 2·6 million wireworms per hectare. While this treated acreage is unnecessarily high in view of the current low numbers present in our grass and arable fields, the cost of the insecticidal seed treatment is very low and poses little risk to wildlife or to beneficial soil animals if used correctly. A suggested compromise is to treat wheat seed for only the first four years after ploughing old grass or a long ley.

Seed treatments containing gamma-HCH give useful control of low to moderate wireworm populations, and are approved for use on both winter and spring wheat. Fonofos seed treatment when used against wheat bulb fly will also give some control of wireworms. However, seed treatments containing chlorfenvinphos or carbophenothion have little effect.

Fonofos granules broadcast and incorporated into the soil at drilling give useful control of both wheat bulb fly and wireworms.

For wireworm populations of 1·25 million or more per hectare, sprays of gamma-HCH or fonofos can be applied overall and worked into the seedbed before drilling.

Wireworm numbers are reduced by birds and many other organisms which feed on them. Culturally, winter or spring wheat should be drilled in a firm seedbed adequately supplied with phosphate. Rolling and top-dressing with nitrogen helps the winter wheat crop to recover from damage.

Chafer beetles

As with wireworms, chafer beetle grub damage to cereals is much less common now than twenty years ago. However, wheat after grass in southern counties of England may occasionally suffer attacks by grubs of the cockchafer, *Melolontha melolontha*, the summer chafer, *Amphimallon solstitialis* or the brown chafer, *Serica brunnea*.

Wheat seedlings are bitten below ground in late autumn. The chafer grub can readily be seen as a dirty white, soft-bodied larva about 3·8 cm long, with a large brown head carrying powerful biting jaws, three pairs of legs and a curved body ending in a swollen tail region.

The grubs take from two to four years, depending on the species, to complete their development. Feeding is confined chiefly to the autumn months. Pupation takes place in the spring and the adult beetles fly actively in the summer, laying their eggs in grassland.

Adequate cultivations following ploughing of infested grassland are sufficient to kill many chafer grubs. Damaged crops should be rolled and given a top-dressing of nitrogen fertiliser to aid recovery.

Cereal leaf beetle (*Oulema melanopa*)

A fuller description of this insect is given in the section on oat pests (see page 219). Damage to winter and spring wheat resembles that in other cereals and grasses, in that long thin strips of leaf tissue are eaten by the beetles and their larvae. The damage is reminiscent of that caused by slugs and the similarity is further heightened by the slug-like appearance of the larva.

Wheat shoot beetle (*Helophorus nubilis*)

Winter wheat after grass leys or herbage seed is sometimes attacked by grubs of the wheat shoot beetle—yet another example of a 'ley pest' which normally lives in grassland but which can transfer to a following cereal crop.

Central shoots or whole plants of winter wheat become yellow from January until late March. A small hole is usually to be seen in the base of an attacked shoot, near which the grub may be found. Each larva is brown and inconspicuous, measuring only 0·6 cm in length. It does not reside and feed within the wheat shoot.

Adult beetles are dull brown and can be found in grass during the summer months. The grubs feed during winter and spring, pupate in April and the adults emerge soon afterwards. There is only one generation a year.

Wheat shoot beetle damage is accentuated when winter wheat is drilled too soon after ploughing infested grass. Early ploughing is preferable to delaying the sowing of winter wheat. Chemical measures are not considered necessary against this pest.

Ground beetles

Most ground beetles are beneficial, feeding on insects (including cereal aphids) and other small animals. One species, *Zabrus tenebrioides*, is an important pest of cereals in some European countries. In England it is not common and only two cases of damage have been recorded; these were to wheat in 1976 at Swaffham Bulbeck (Cambs) and to barley in Oxfordshire during 1983. In both cases the beetle larvae shredded the leaves and reduced the plants to a fine fibrous mat lying at first on the soil surface, then subsequently dragged below ground by the burrowing larvae. Damage to the developing grain, common under continental conditions, was not observed in these cases.

Cereal flea beetles

A number of flea beetle species feed either on a variety of cereals, e.g. barley flea beetle on wheat, barley, oats, rye and maize, or feed exclusively on one cereal, e.g. wheat flea beetle on wheat. The adult beetles are small and have well-developed hind legs which enable them to jump vigorously.

The wheat flea beetle (*Crepidodera ferruginea*) is a typical 'ley pest'. Adult beetles are reddish and are present from June until September. They lay eggs chiefly near grass plants. The grubs are creamy white with a black head, small black legs and a dark plate near the tail. They normally feed within grass shoots during winter and spring. Pupation takes about four weeks.

If infested grass is ploughed and followed shortly afterwards by winter wheat, the grubs migrate upwards from the rotting turfs and enter young wheat shoots. A small, neat hole at the base of the shoot with the tiny grub inside the yellowing central tissue will confirm the cause of trouble, although the larvae are often to be found in the soil around the damaged plant.

Control depends on leaving a gap of at least four weeks between ploughing grass and drilling a following winter wheat crop. Chemical control measures have not been developed. Damage by this pest alone is rarely sufficient to warrant any action.

Sawflies

This group of insects is closely allied to bees, wasps and ants. The only species of interest as a pest of wheat is the wheat stem sawfly (*Cephus pygmaeus*). This is important in the wheat lands of North

America and elsewhere, but nowadays it is rarely recorded in the United Kingdom.

Damage occurs to winter wheat and (less commonly) to spring wheat. The larva feeds downwards within the wheat stem until it gets to about 2·5 cm above ground level. The stem may break off at this point in rough weather. Close examination of the broken end of the stubble shows a plug of sawdust-like material with the whitish maggot measuring about 1·2 cm in length within. It has a brown head capsule and no legs and spends the winter within a cocoon in the stubble. The larva pupates in May and emerges soon afterwards as an adult sawfly, with black and yellow body markings and two pairs of wings. After mating, the female sawfly lays her eggs in a slit made in the wheat stem just below the ear.

The practice of burning wheat stubbles after harvest has probably contributed to the decline in importance of this pest, and special control measures are not required.

Aphids (plant lice) and leafhoppers

This group of insects includes a number of important pests of wheat. They possess piercing and sucking mouthparts which are used to obtain juices from various parts of the plant. In addition to direct injury resulting from their feeding activities, aphids may transmit barley yellow dwarf virus (page 200) from infected to healthy plants. The leafhopper *Javesella pellucida* transmits a virus-like mycoplasma disease, the causal agent of European wheat striate mosaic (page 135), which is fortunately rare in Britain.

Damage symptoms

Direct feeding injury usually depends on the presence of large numbers of aphids on the plant. Stems and foliage may be stunted, with yellow, reddish or purple spots on the leaves marking the feeding sites. Leaf tips may turn brown and die. The symptoms vary according to which aphid species is present.

Leafhoppers are rarely present in large numbers and a few pale spots on the leaves are the only visible sign of their activity.

The diseases are described elsewhere (page 134). We are concerned below with the aphids themselves, their description, direct injury and control.

Bird-cherry aphid (*Rhopalosiphum padi*)

This is a small brown or greenish-brown insect found in late spring on the lower leaf sheaths of winter wheat or in late autumn or even during mild winters in sheltered parts of fields in southern England. The aphid does not usually multiply rapidly on wheat, and rarely occurs in large numbers on that cereal during the summer months. In northern counties of England and in Scotland the aphid does not usually survive the winter on cereals or grasses, and migrates in autumn to the bird-cherry (*Prunus padus*), on which it lays its eggs in crevices in the bark. Further south in England and Wales, the aphid is able to continue living on cereals and grasses at least in mild winters.

The chief importance of *R. padi* lies in its ability to transmit strains of barley yellow dwarf virus (page 200). Direct injury to wheat is a rare occurrence.

Apple-grass aphid (*Rhopalosiphum insertum*)
This aphid spends the winter in the egg stage on apple, pear, hawthorn and other woody rosaceous plants. It migrates in late spring to wheat, barley, oats and grasses, on which it colonises the roots and underground stems. This aphid seems to be unimportant on cereals despite its ability to transmit barley yellow dwarf virus.

Fescue aphid (*Metopolophium festucae*)
The fescue or grass aphid is a light green insect which spends the winter on grasses either in the egg stage or (in mild seasons) as a live aphid. Its numbers often increase after mild, open winters and the resulting migration to grasses and cereals can spell trouble. Feeding symptoms on cereal leaves are of red or purplish discrete spots. This aphid was fairly common in 1971.

Rose-grain aphid (*Metopolophium dirhodum*)
This is a moderately sized aphid, light green with a dark green stripe running centrally along the back. It spends the winter as an egg on roses. Winged aphids migrate to grasses and cereals in late spring and their wingless progeny can usually be found towards the centre of wheat fields. In years such as 1970 and 1979 this species can build up rapidly so that the whole lower surface, and occasionally the upper surface, of each leaf and the stem may be covered with tightly packed bugs. The leaves then become coated with honey-dew, a sticky excrement produced by the aphids.

The rose-grain aphid is capable of transmitting strains of barley yellow dwarf virus. Its importance as a direct pest of wheat is less than that of the grain aphid and depends on the extent to which it colonises the flag and upper leaf. *M. dirhodum* is usually found only on the lower leaves, which turn yellow and senesce prematurely. Losses in grain yield of 7–10 per cent have been recorded in recent trials but further work on both quantitative and qualitative losses attributable to this aphid are needed.

Grain aphid (*Sitobion avenae*) (Plate 15)
The grain aphid is variable in colour and can be black, brown, green or pink. Eggs are laid on grasses, and in late spring winged forms move to cereals and grasses. As with other migrating cereal aphids, *S. avenae* is earlier colonising cereal crops in the south and west of England than further north.

It usually colonises cereal plants on the edge of a field, but under favourable conditions wingless aphids can soon be found nearer the centre of the crop. The first colonies inhabit the upper leaves of winter and spring wheat, but as the flowering heads emerge the aphids move on to the stem and subsequently to the ears. At this stage infestations of the reddish aphids are easily spotted by the farmer. As the grain hardens the cereal plant becomes less acceptable to the aphid, which migrates in large numbers to its winter hosts (grasses) usually at the end of July.

The grain aphid transmits strains of barley yellow dwarf virus (page 200). Its direct harm to wheat is greatest when infestations develop on the ears or at just before the start of flowering. Copious quantities of honeydew can be exuded on which sooty moulds (page 145) quickly develop, giving the heads a dirty appearance. Infestations which develop much later than the flowering stage are of much less consequence economically.

Recent experimental work in England and other countries of north-west Europe suggest that direct feeding injury arising from grain aphid attacks on the ears can mean a loss of 12 per cent in grain yield in both winter and spring wheat, while flour quality can be adversely affected in bread-making wheat varieties. The aphids intercept nitrogen assimilates intended for the developing grain. Aphids feeding on the leaves probably cause less damage.

Blackberry aphid (*Sitobion fragariae*)
This is a dull green aphid, similar in size to *S. avenae*. It migrates from bramble, the overwintering host, to wheat and other cereals in June and July and forms colonies on both leaves and ears. Its numbers rarely approach those of the grain aphid and little is known of its economic status.

Plate 15. Grain aphids, *Sitobion avenae*, (×6) infesting wheat ear.
Farming Press.

Cereal leaf aphid (*Rhopalosiphum maidis*)
This is a bluish species commonly found on maize, spring barley and winter barley volunteer plants. Its numbers are rarely large on these hosts and its economic status is uncertain.

Other aphid species may be found as yellowish or pink, rounded, waxy insects on the roots of wheat and other cereals. They are probably of no economic importance.

Control of cereal aphids
Control measures are directed against (a) the aphid vectors of barley yellow dwarf virus (page 200) or (b) aphids causing direct injury to cereals.

Direct injury is usually caused by summer infestations of the grain aphid, rose-grain aphid or grass aphid. Attacks are usually seen in central and southern areas of England only. For the grain aphid, an average of five or more aphids per ear on winter wheat at the start of or during the flowering period, when the weather is warm and settled and favourable for an increase in aphid numbers, justifies the application of a single spray of an approved aphicide. Limited experimental evidence suggests that spray treatment at or just after ear emergence may give a greater yield response. Treatments applied even earlier may have to be repeated if aphids recolonise the crop. Sprays applied after the flowering period and towards grain ripening give progressively smaller yield increases. Recent work has demonstrated worthwhile yield increases from single aphicidal sprays applied up to the milky ripe or early dough stages of grain ripening if thirty or more rose-grain aphids are present on the flag leaf.

Most aphicides are equally effective whether applied by ground sprayers or from the air. An aphicide with low toxicity to non-target organisms is preferred. Under epidemic conditions, the call for aircraft is so heavy that many crops are sprayed too late.

Farmers interested in maximising wheat yields often apply aphicides as routine insurance treatments. Such indiscriminate use could quickly lead to the development of resistance to insecticides, and cannot be justified on economic grounds and because of possible risks to wildlife. Furthermore, the unnecessary early application of broad spectrum insecticides often leads to a resurgence of aphid populations as natural enemies are killed. An additional treatment may then be required.

Cereal aphids are subject to predation by ladybirds, hover fly larvae, ground beetles and earwigs, and parasitised dead aphid bodies ('mummies') can often be found in large numbers on the host crop. It is becoming clear that these natural enemies play an important role in regulating cereal aphid numbers.

As far as cultural control is concerned, the importance of completely burying turf when ploughing grass for autumn cereals should be mentioned. Many cases of virus transmission from grass to winter cereals have occurred, particularly in western counties, because viruliferous aphids were able to survive on the ploughed grass and migrate directly to the cereal crop when it emerged (see page 202).

Thrips

Thrips or thunder-flies are small brown or black insects, equipped with piercing and sucking mouthparts and usually having two pairs of narrow wings. They are particularly common on hot, sunny days in summer.

Leaves, stems and ears of wheat are attacked, the adult thrips and their immature stages sucking sap from the outer tissues and leaving silvery marks as evidence of their feeding. When numbers are particularly large, shrivelling of the grains may result.

The grain thrips (*Limothrips cerealium*) is commonly found on wheat and other cereals. The adults are black or brown and only the females have wings. The young stages (nymphs) are orange-yellow. Eggs are laid in spring within cereal plant tissues and the thrips feed among folded leaves and later within the ears. The adult thrips spend the winter in hedge bottoms and similar habitats.

Little is known of the precise importance of thrips feeding and early accounts of damage to shoots and ear and 'blindness' of florets resulting from thrips attack can probably be discounted. Control measures for cereals have not been advocated. However, in 1978 thrips damage to the lemna and palea of developing spikelets of wheat was confirmed in Yorkshire and occasionally such damage appeared to be associate with a subsequent disease of the spikelets caused by *Botrytis* (see page 145). Thrips were also found feeding on the germ, especially of barley, and this was tentatively associated with reduced or delayed germination of the grain—an important factor where malting barley is concerned.

Several other species of thrips can often be collected from cereals, either feeding or merely sheltering in the crop.

MITES

Mites, together with spiders, scorpions etc., form a group allied to insects, but from which they differ in possessing four pairs of jointed legs. The few species found on cereals in this country are extremely small and need to be identified by an expert.

Oat spiral mite (*Steneotarsonemus spirifex*)
This is more commonly found on oats (page 220) and several grasses. It occasionally attacks wheat, the flowering stem often being trapped within the flag leaf and failing to emerge; or when it emerges is reduced in size.

Grass and cereal mite (*Siteroptes graminum*)
This mite causes 'silver-top' in wheat, barley, oats and certain grasses. The flowering heads turn white and bleached and can be easily pulled away from the leaf sheath. Close examination of the cut stem surface often reveals several spherical, glistening white female mites. The same species is associated with a bud rot of carnations.

Mite problems in our cereal crops are not sufficiently important to warrant control measures.

MILLEPEDES (Plates 16 and 17)

These soil-dwelling animals are distantly related to insects but are clearly different in having numerous pairs of legs. It is important to distinguish between millepedes and centipedes, because the latter are useful to the farmer and devour wireworms and other harmful organisms. Each segment of the millipede body carries two pairs of legs, whereas in centipedes each body segment bears only one pair of legs.

Millepedes have small biting mouthparts which enable them to feed on living parts as well as on decaying vegetable matter. They can be conveniently divided into two groups—those with cylindrical bodies and those with flattened bodies. The first group contains the spotted millepede, *Blaniulus guttulatus*, and the black millepede, *Cylindroiulus londinensis*. Flat millepedes include *Polydesmus angustus* and *Brachydesmus superus*. All four species have been implicated in occasional reports of damage to cereals, shortly after germination, when the plumule is nibbled or seedling stems cut clean through. In the latter case several adjacent plants may be affected. Millepedes are more often found feeding secondarily on cereal tissues which have already been damaged by other pests.

Control measures have not been advocated.

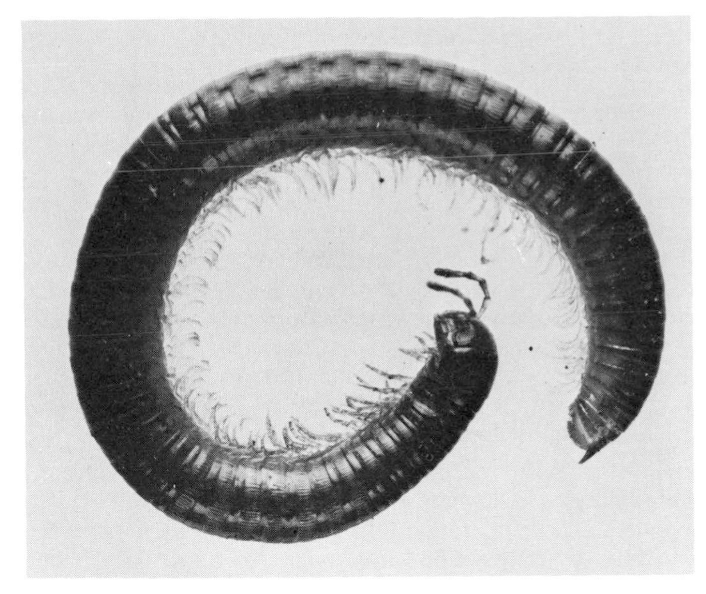

Plate 16. Black millepede (×4).

Plate 17. Spotted millepede (×14). Note double pair of legs for each body segment.

SLUGS (Plates 18 and 19)

Slugs are land molluscs which are characterised by a soft, slimy legless body, pointed at both ends with the head end bearing two pairs of tentacles on one pair of which the eyes are situated. Slugs differ from their close relatives the snails in having a greatly reduced shell which is rarely visible and which usually lies beneath the body surface.

The species of greatest importance in damaging wheat and other cereals is the field slug (*Deroceras reticulatum*) to which the following account largely refers. Other species such as *Deroceras laeve* and the garden slug (*Arion hortensis*) may occasionally be important.

Damage symptoms

Winter wheat is damaged at three growth stages by slug grazing. The seed may be hollowed out shortly after being drilled, a ragged hole indicating where feeding has occurred. Other pests such as bibionid fly larvae (page 48) may cause similar holing of the grain but slimy trails in the soil betray the presence of large numbers of slugs, and microscopic examination of the grain reveals the characteristic marks left by the slug's rasping organ. Next, the young shoot may be completely bitten off below ground, leaving a ragged cut end. In both cases, complete loss of plant may result from slug feeding and the wheat fails to establish. Finally, the young seedling leaves may be shredded and torn above ground, with long thin windows left in the frayed leaf tissue.

Life history

The field slug is about 5 cm when full grown and is variable in colour, ranging from grey to brown or even reddish with a dark mottled appearance. It is more of a surface feeder than other slug species and can readily be found in mild, damp weather on or just below the soil surface or climbing plant foliage. When soil conditions are dry and temperatures are either high or very low, adult slugs move deeper into the soil. Severe frosts can decimate numbers of young slugs.

Feeding and movement at the soil surface usually occur at night. The field slug exhibits a form of 'homing' behaviour whereby individuals often return to the same point from which they started the night's wanderings.

Each slug is a hermaphrodite, possessing male and female sex organs. Packets of sperms are exchanged after an elaborate court-

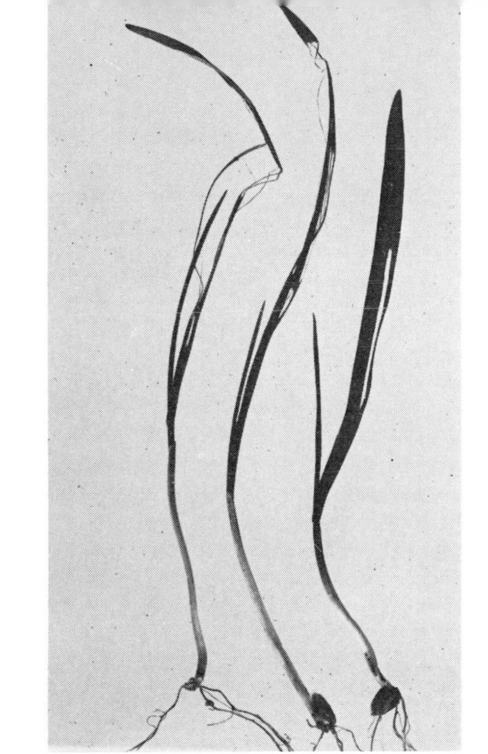

Plate 18. Slug damage to autumn-sown wheat seedlings, showing characteristic shredding of leaves.
Crown copyright.

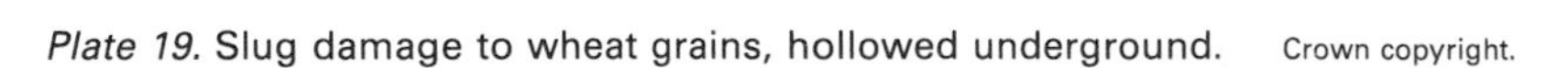

Plate 19. Slug damage to wheat grains, hollowed underground. Crown copyright.

ship procedure, which can occur at any time of the year. Eggs are laid in the soil in groups of up to fifty. Each egg is spherical and translucent grey in colour. The time taken for the egg to hatch varies from three weeks to four months according to soil temperature. The baby slugs resemble adults except for their smaller size and lighter colour. They commence feeding immediately on a wide variety of plant tissues. The time taken for a juvenile slug to become sexually mature varies greatly.

Peak numbers of the field slug are usually recorded in late spring and autumn, suggesting two generations completed in each year although the pattern is probably more complex than this. The effects of weather, cropping, cultivations and predation affect the life cycle and survival prospects. Minor peaks in numbers at other times of the year suggest that under favourable conditions extra generations can be completed.

Control

The influence of such predators as birds, mammals, toads and ground beetles has probably been underestimated and in addition there are parasites of slugs including nematodes and the larvae of a fly. Weather plays an important role in regulating slug numbers, but more significantly affects slug activity at the soil surface.

The field slug is most common, and its attacks on winter wheat are most frequently recorded, on heavy clays, clay loams and medium to heavy silts. Numbers are far fewer on light-textured soils and damage is rarely seen in fen peats.

Serious damage often occurs in direct-drilled winter wheat and in winter wheat following clover, grass, oilseed rape, peas and beans; these crops, and the trash left on the soil surface following their harvest, provide a suitable environment in which slugs can feed and multiply.

Well-consolidated soil restricts slug movement, and wheat on the headlands often remains undamaged while the centre of the field is destroyed. The traditional tilth for winter wheat on heavy soils is a rough and cloddy one—conditions which favour slugs. Where damage is anticipated, cultivations before sowing winter wheat should aim at producing a well-consolidated soil with as fine a top tilth as possible. Early shallow drilling is to be recommended. Spring wheat should not be drilled too early on slug-infested land.

Attempts at predicting the likelihood of damage to winter wheat crops have not as yet proved reliable because they depend on estimates of slug activity at the soil surface. A change in weather conditions between the activity estimate and sowing time can seriously affect the accuracy of the prediction.

The ideal chemical control measure is one which protects the wheat seed from sowing until shortly after emergence, since the late grazing of the seedling leaves is not usually very important. A molluscicidal seed treatment would seem to fit theoretical requirements, but many of the most active chemicals tested as seed treatments have been either too phytotoxic or are too great a risk to grain-eating birds.

At present chemical control relies on the use of surface attractant baits containing methiocarb or metaldehyde broadcast over the soil surface just before or a few days after drilling, or combine-drilled with the seed. Such baits are only partially effective because they affect only that proportion of the slug population which comes to feed on the soil surface within a few days of baiting. Baits usually deteriorate quickly because of fungal contamination or lack of weather-proofing. Metaldehyde immobilises slugs for several hours, but in suitably damp conditions they can recover. Slugs do not recover from methiocarb poisoning, and this material is preferred when moist weather conditions prevail at the time of baiting. Other proprietary formulations of inorganic salts are available but attempts to assess their performance on cereals have as yet been inconclusive.

NEMATODES (EELWORMS)

These animals are mostly worm-like in shape and often too small to be seen with the naked eye. They live in moisture films in the soil, in fresh or marine water, or are parasitic on or within plant and animal tissues. Plant-parasitic nematodes all have a head spear used in feeding to penetrate cellular tissues. Many kinds attack cereals but we shall mention only those believed to be of some consequence.

Wheat gall nematode (*Anguina tritici*) (Plates 20A, B)
This is the nematode responsible for forming ear cockles or galls in wheat and rye heads. It deserves only a mention because it has now become virtually absent in Britain, thanks to the efficiency of modern methods of seed cleaning and to the fact that larvae in the soil cannot survive long in the absence of a host crop.

Infested wheat heads have one or more grains replaced by brown or black rounded galls, each gall containing a mass of larval nematodes which are able to withstand drying. When an ear

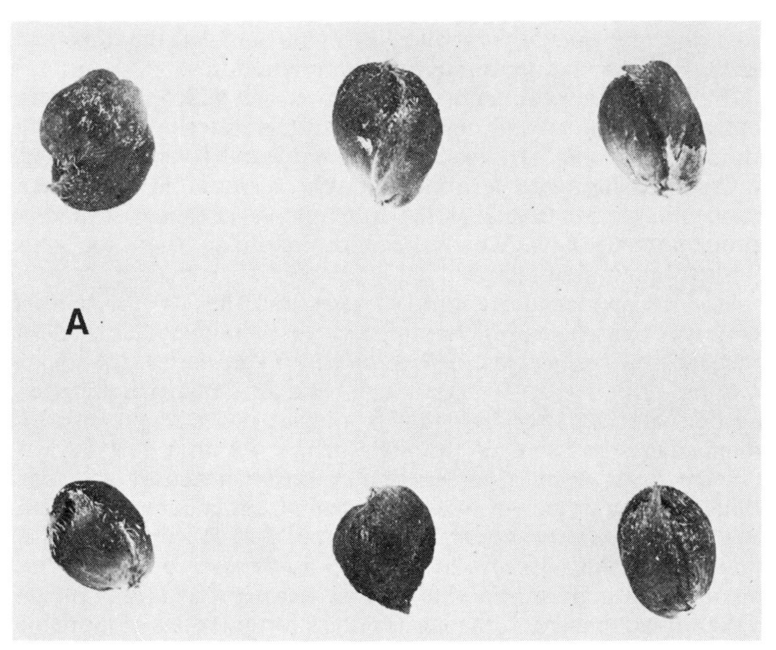

Plates 20A, B. Wheat gall nematode. (a) Ear cockles of wheat compared with (b) normal wheat grain. Crown copyright.

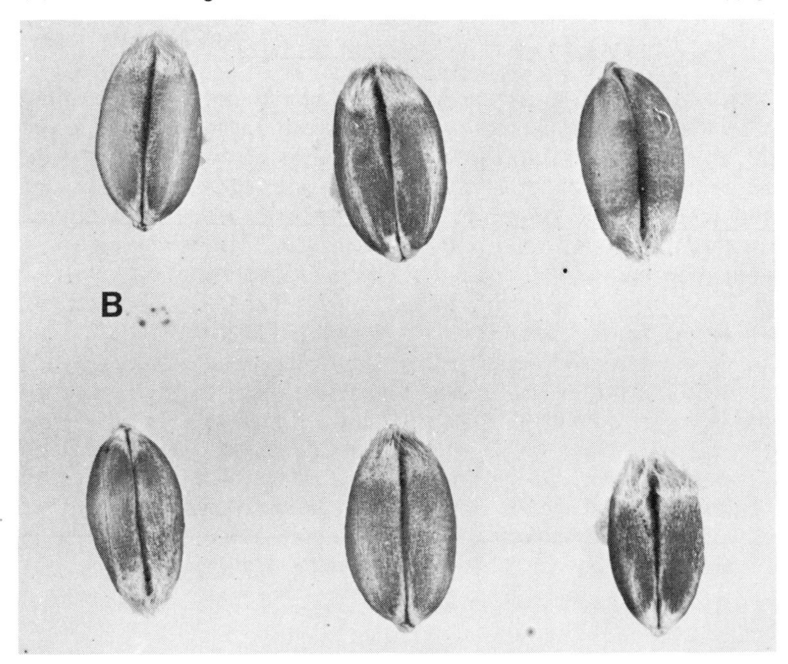

cockle is sown in the soil with healthy grain, it breaks down to release the nematode larvae. These invade host plants to live between the leaves, and are later carried upwards on the stem and invade the flowers. There the nematodes mature and mate. The eggs soon hatch but the larvae remain dormant in the gall and can remain viable for several years. Rye is similarly attacked.

Root-ectoparasitic nematodes

Ectoparasitic nematodes feed on the outside of plant tissues. A few genera are occasionally able to penetrate and feed within the host cells. Some root-feeding forms are extremely common in cereal fields, and they include spiral nematodes (*Helicotylenchus* and *Rotylenchus*), stunt nematodes (*Tylenchorhynchus*, *Merlinius* and *Amplimerlinius*) and ring nematodes (*Criconemoides*). They have been shown to cause serious damage to a wide range of crop plants in warmer climates, but seem to be less important in British soils. Spiral nematodes are thought to contribute to 'winter kill' of winter wheat on the Yorkshire Wolds.

Other ectoparastic nematodes are important not only for causing direct injury to roots but also because they transmit virus diseases from infected to healthy plants. No cereal virus has yet been implicated in such nematode-vector activity in the United Kingdom.

The stubby-root nematodes, *Trichodorus* and *Paratrichodorus*, are so called because their feeding at the root tips lead to browning and thickening of the roots, with large numbers of lateral rootlets formed which may in turn be attacked. These nematodes are responsible for 'docking disorder' of sugar beet on light sandy soils. Winter wheat on such soils may be stunted, but effects on yield are as yet unknown. Low rates of organophosphate or oximecarbamate nematicide granules have given peak yield responses of 1 t/ha in winter wheat in soils heavily infested by trichodorid nematodes. *P. anemones* has been shown recently to cause stunting and loss of yield in spring wheat when present at initial densities of about 300 per litre of soil.

The needle nematode, *Longidorus*, is another nematode associated with 'docking disorder' of sugar beet. Its feeding produces small, often terminal, swellings on roots. Numbers are particularly high in certain sandy and peat soils in Shropshire and in the Vale of York and recent work has established a relationship between *L. elongatus* numbers and yield and size of spring barley grain. The influence (if any) of *Longidorus* on growth and yield of wheat, rye and oats has not yet been determined.

The dagger nematode, *Xiphinema*, affects root systems in a similar manner to *Longidorus*, but we know nothing of its status as a pest of wheat and other cereals. It is not common in arable soils in Great Britain.

Root-lesion nematodes (*Pratylenchus* spp.)
Several species of root-lesion nematodes live in British soils. A few feed externally on roots of wheat and other plants, but most invade the root tissues and feed as endoparasites, causing cavities and lesions through which secondary fungi and bacteria may enter the plant. While many species of *Pratylenchus* are undoubtedly pathogenic in wheat and other cereals, their effects on cereal growth and yield are largely unknown. One species, *P. fallax*, has been shown to stunt spring barley in fields on the Bunter sandstone of Nottinghamshire, whereas another (*P. neglectus*), even when present in large numbers, appears to be of little consequence in cereals in East Anglian alkaline soils. On the continent of Europe, damage caused by these nematodes is usually associated with soil acidity problems.

No practical control measures are known.

Cereal root-knot nematode (*Meloidogyne naasi*) (Plate 21)
This nematode is known chiefly in South Wales and south-west England as an endoparasite of wheat, barley and grasses. It is a pest of sugar beet in Belgium, of spring wheat and barley in the Netherlands and barley in California.

All cultivars of winter and spring wheat seem to be efficient hosts of the nematode. Thickened roots or roots with clubbed tips or with swellings along their length indicate infestation. Affected crops show areas of stunted plants from late spring onwards. Older leaves are yellowed, the plants produce fewer tillers and fewer or smaller grains. Seedlings may be killed outright by this nematode.

In many infested fields, other factors such as soil-borne diseases, poor drainage or nutritional deficiencies may contribute to the patchiness of the crop and confound the symptoms of nematode attack.

For further information on life history of the pest, yield losses in barley and on possible control measures, see page 167.

Cereal cyst nematode (*Heterodera avenae*)
The life history, occurrence and control of the cereal cyst nematode are dealt with under pests of oats (page 221).

Wheat plants which are heavily infested by this endoparasite

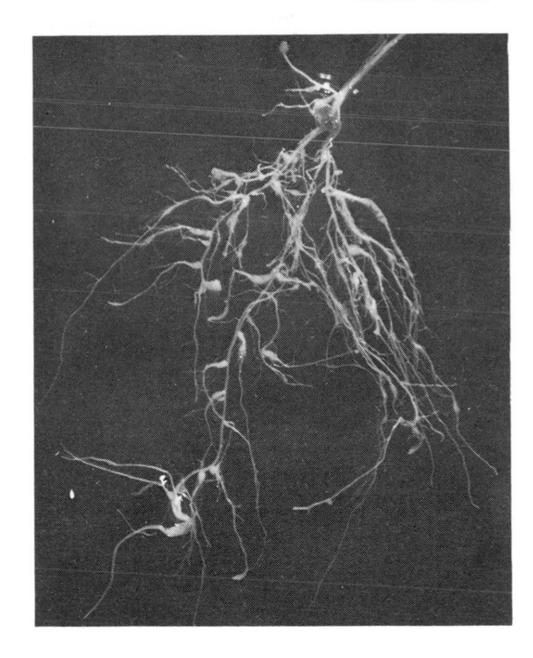

Plate 21. Cereal root-knot nematode. Roots of infested winter wheat seedling showing sickle-shaped and terminal galls.

show branching and knotting of the shallow root system, with white cysts appearing on the roots from June until harvest. The plants may be stunted but heavily infested patches more usually have a pale, sickly appearance. Spring wheat suffers more damage than winter wheat, and yield losses of 188 kg/ha for every increase of ten eggs per gram of soil have been recorded in spring wheat. Treatment with an effective soil nematicide such as dazomet has given yield increases of up to 1253 kg/ha in Kloka spring wheat, although part of this yield increase is attributable to side effects of the nematicide.

At least one cultivar of spring wheat—Loros—is known to be partially resistant to UK pathotypes of cereal cyst nematode, its resistance being controlled by a single dominant gene. Little use has yet been made of genetic sources of resistance in breeding programmes for spring wheat.

BIRDS (Plates 22 and 23)

Seed wheat left lying on or near the soil surface is often devoured by pigeons and pheasants. Seed treated with chemicals and especially with insecticides for wheat bulb fly control should there-

Plate 22. Birds. Ears of barley stripped by house sparrows. Arable Farming.

Plate 23. Birds. Wheat leaves pecked by moorhens. Crown copyright.

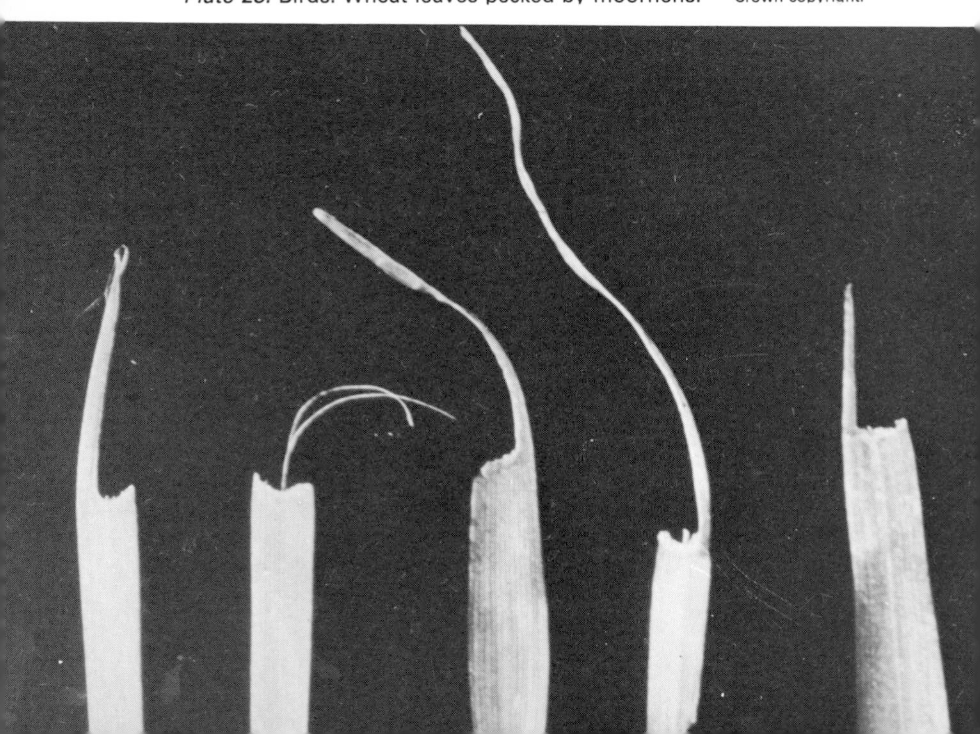

fore not be left as 'spillages', for some of these materials are toxic to grain-eating birds. Seed drilled at the normal depth may be hooked out by rooks, starlings, pheasants or jackdaws, and long lengths of row may be entirely removed.

At or shortly after emergence, the young cereal leaves and shoots are the target for attacks by flocks of skylarks or finches. The plants are cut off just above soil level and left stewn over the ground. Triangular beak marks may be discernible on the cut tissue, but more commonly a V-shaped notch or diagonal cut end serves to distinguish bird damage from that caused by mammals, in which the cut runs straight across the shoot or leaf.

Tillering cereal plants are occasionally pulled out of the ground by rooks, starlings, or jackdaws; the plants are left lying on the soil surface which may be pitted by holes made by the beak probes. Pigeons often seen in cereal crops are probably feeding on young charlock and not on the leaves of cereal plants.

In most years since 1972, brent geese have grazed cereals in coastal fields of East Anglia and Sussex in the period December to March. Paddling of the geese often leads to soil capping problems.

Bird damage decreases as tillering ends in early summer, although geese and swans may continue to feed near large waterways. Moorhen damage may be evident on the edges of fields close to ponds and dykes.

The ripening grain attracts such birds as sparrows, rooks, pheasants, pigeons, ducks and geese. Flocks of sparrows concentrate on headlands and lodged crops, when flowering heads may be completely stripped of the grain. Larger birds may peck at the straws and cause the heads to fall to the ground. Lodged areas of wheat fields are particularly liable to attack.

MAMMALS

Domesticated animals such as sheep, goats and cows occasionally stray in cereal fields and cause some damage. Wild or feral mammals can sometimes cause serious losses to individual cereal crops.

Seed corn can be dug out of the drill by rats and, less frequently, by field mice. Rats and voles tend to graze only on the edge of a cereal crop, and they often tunnel into the soil. Rat damage typically consists of a shallow channel cut along the drill with cereal stems often deposited at one end.

Moles may push up young seedlings, but the main damage to tillering cereal plants is that caused by grazing of rabbits, hares, voles and field mice. The chief periods of grazing are in late autumn, winter and early spring. The cut edge of a leaf grazed by a mammal usually runs straight across, i.e. at right angles to the long axis of the leaf.

Rabbit damage is becoming increasingly common and in some areas is approaching the pre-myxomatosis levels. Patches in the field may be heavily grazed, especially along headlands in the vicinity of hedges or copses, or the whole cereal crop can be evenly grazed. Rabbit runs and droppings are usually much in evidence.

Hare damage more typically occurs in open country. Short lengths of row are grazed, and the droppings differ from those of rabbits in being larger and somewhat flattened.

Winter wheat may be grazed by field mice, especially in years when these small mammals are numerous. Grazing starts close to the edge of a field nearest to a copse or rough land, then advances along a broad front into the field. Field mice often collect cereal leaves and place them in heaps around the edge of the field.

Deer may graze cereal fields, especially in prolonged spells of hard wintry weather although more serious damage results from browsing crops in ear. The hoof-marks and long, pelleted droppings can usually be found as evidence of deer feeding.

As with birds, attacks by mammals decrease as cereals reach the end of tillering. Squirrels, rats, mice, voles, rabbits and hares may bite through occasional stems. The same species are responsible for damage near to harvest, when the stems may be gnawed and the fallen heads scuffed for ripe grain. Similar drainage has been ascribed to coypus feeding along the edges of cereal crops close to water courses in Norfolk and Suffolk.

Chapter 3

WHEAT DISEASES

PLANT DAMAGE SYMPTOMS

Disorder	*Cause*	*Symptom*	*Page*
SEEDLINGS			
Failure to emerge	Organomercury injury	Seed does not germinate or seedlings with short thickened shoots, stunted roots.	23
Failure to emerge	Gamma-HCH injury	Similar to and often associated with organomercury injury but seedling roots and shoots are club-shaped.	23
Delayed emergence	Deep drilling	Seedlings yellowish and delayed sometimes in patches.	84
Delayed emergence	Blue mould	Blue mould on seed grain.	84
'Rugby stocking'	Deep drilling	Seedlings with distinct discoloured bands on newly-emerged first leaf.	84
Seedling blight	*Fusarium* spp., *Septoria nodorum*	Seedlings small, stunted, discoloured and may be killed soon after emergence.	95 125
Browning root rot	*Pythium* sp.	Patches of yellowed seedlings, roots brown.	85
Root rot	*Rhizoctonia solani*	Poor growth, sometimes purple, in distinct patches.	85
Snow mould	*Fusarium nivale*	Patches of yellow plants with pink fungal mould; especially after snow.	95
Snow rot	*Typhula incarnata*	After snow, patches of yellowed or rotted	86

Disorder	*Cause*	*Symptom*	*Page*
		plants with small brownish round fungal bodies on dead tissue.	
Frost lift	Winter frost	Small or large patches of sickly plants in spring.	86
May yellows	Temporary nitrogen deficiency	Extensive areas, usually whole fields yellow.	175

ROOTS AND STEM BASE

Take-all	*Gaeumannomyces graminis*	Roots, and later some stem bases, blackened; can cause stunting and premature death of plants usually in patches, occasionally at random.	87
Brown foot rot	*Fusarium* spp.	Stem bases brown, later with pinkish spore masses mainly on joints; stems break easily at node just above soil level; mainly at random, occasionally patchy.	95
Eyespot	*Pseudocercosporella herpotrichoides*	Eyespot lesions on stem bases; stems may twist or break at lesion causing crop lodging.	98
Sharp eyespot	*Rhizoctonia cerealis*	Lesions more numerous, higher up stem and more clearly defined than for eyespot; later sometimes large lesions with 'water marks' on sheaths and white mould between stem and sheath. Rarely causes crop lodging.	106
Foot rot	*Cochliobolus sativus*	Rot at stem bases; uncommon.	109

STEM AND LEAF

Yellow rust	*Puccinia striiformis*	Bright orange-yellow pustules arranged in	109

Disorder	*Cause*	*Symptom*	*Page*
		lines; pale green or brown stripes on leaf blade.	
Brown rust	*Puccinia recondita*	Brown pustules scattered at random on leaf.	114
Black stem rust	*Puccinia graminis*	Orange-brown (later black) pustules mainly on sheath and stem.	116
Mildew	*Erysiphe graminis*	White fluffy pustules on leaf and sheath.	118
Leaf spot	*Septoria nodorum*	Pale brown isolated oval spots, later large irregular areas or small brown spots on upper leaves and sheaths. Brown spore cases (pycnidia) sometimes on affected areas.	125
Speckled leaf spot	*Septoria tritici*	On young plants pale spots on extensive areas. Later on upper leaves rectangular lesions, with yellow halos, may merge. Black pycnidia usually obvious on all lesions.	128
Late leaf disease	Various fungi	Brown lesions or speckling often extensive on flag leaf or other upper leaves after flowering.	132
Cephalosporium stripe	*Cephalosporium gramineum*	Broad yellow stripes on leaves; mainly isolated plants.	133
Barley yellow dwarf	Barley yellow dwarf virus	Upper leaves especially flag leaf with pale yellowing from the tip. Autumn-sown crops may have patches of stunted plants.	134
European wheat striate mosaic	(Mycoplasma)	Yellow-white narrow stripes on leaves especially in young plants; isolated plants	135

Disorder	*Cause*	*Symptom*	*Page*
		or patches may be killed.	
Manganese deficiency		Leaves pale green with very small pale brown spots at late tillering stages.	135
Copper deficiency		Young leaves tipped, twisted; sometimes dark discolouration of stems.	149
Leaf tipping	Late frost damage	Leaf tips white-pale brown to a line across leaf; usually 2–3 successive leaves affected.	148
EARS			
Bunt	*Tilletia caries*	Glumes spreading; inside of grain replaced by black spores, with fishy smell. Spores not shed in the field.	136
Loose smut	*Ustilago nuda*	All grains replaced by a mass of black spores; obvious at emergence; spores are shed leaving a bare stem.	138
Rusts	*Puccinia* spp.	Yellow-orange or brown, later black, pustules; see leaf symptoms above.	
Mildew	*Erysiphe graminis*	White, later brown, superficial mould mainly on glumes; black spore cases often present.	118
Glume blotch	*Septoria nodorum*	In green ears purple brown blotches near tips of glumes; in ripened ears blotches less well defined, whitish sheen, black spore cases embedded in tissues.	125
Ergot	*Claviceps purpurea*	Hard purple-black bodies replacing grain	140

Disorder	*Cause*	*Symptom*	*Page*
		and protruding from a few spikelets.	
Scab	*Gibberella zeae*	Pink-red fungus with black bodies embedded on surface of ear.	144
Ear blight	*Fusarium* spp.	Individual spikelets or entire parts above affected stem bleached. Salmon pink spore masses usually present.	97
Botrytis	*Botrytis cinerea*	Individual spikelets at random; grain destroyed, large lesion on glume. Grey mould.	145
Twist	*Dilophospora alopecuri*	Ear infected in sheath, covered with white, later black, mould; rare.	145
Black mould	*Cladosporium herbarum* and others	'Sooty' black moulds on surface; ears often thin.	145
'Whiteheads'	*Gaeumannomyces graminis* *Pseudocercosporella herpotrichoides* *Rhizoctonia cerealis* *Fusarium spp.* *Cephalosporium spp.* Late frost damage Loose ear Herbicide injury etc.	Ears bleached, prematurely ripe, grain shrivelled; a symptom of many disorders (see also roots and stem base diseases).	
'Blindness'	Frost damage	Many or few spikelets bleached and blind in groups on one part of ear, frequently at the tip.	148
Copper deficiency		Terminal spikelets blind, or 'rat-tailed'; grain shrivelled, delayed ripening.	149
Loose ear		Ear and stem below bleached; rest of plant green, stem severed in flag leaf sheath.	150
Distortion	Herbicide injury	Spikelets opposite, more numerous or bunched, etc., ear trapped in sheath.	150

Disorder	*Cause*	*Symptom*	*Page*
GRAINS			
Bunt	*Tilletia caries*	Grain filled with masses of black spores; fishy smell.	136
Black point	*Alternaria* spp.	Brown areas at embryo end of grain.	152

Deep drilling

Drilling too deeply because the drill is set incorrectly or because of uneven land can lead to delayed and sometimes uneven emergence and to a yellowing of the seedlings. Under these conditions seedlings are more susceptible to soil-borne fungal diseases (e.g. those caused by *Fusarium*, *Rhizoctonia*, *Pythium*), especially when growing conditions are unfavourable. Once the emerged seedlings are established, subsequent growth is normal and the crop then appears to be quite unaffected.

A colour banding or 'rugby stocking' symptom on leaves of wheat and oat seedlings at emergence has also been associated with excessively deep drilling, especially where the soil is heavily consolidated after drilling. The symptoms disappear a week or so after emergence.

Blue mould on seed (*Pencillium* spp.)

Occasionally a blue mould can be found on ungerminated seed and on the remains of seed attached to seedlings. The condition is usually noticed when seeds are examined in crops showing delayed or irregular emergence.

The blue mould, which is a species of *Penicillium*, is very prominent and appears most commonly on seeds in soils which have remained very dry after drilling and where germination and subsequent growth have been slow. Under such conditions food reserves in the seed are not used by the developing seedlings as quickly as they would be under more favourable soil conditions, and the mould occurs along lines or cracks in the seed coat where the food reserves leak from the seed. Although the fungus is not regarded as a pathogen of wheat, under the conditions described it probably acts as a competitor with the seedling for food reserves in the seed. Such seedlings may be retarded and fail to establish.

The growth of *Penicillium* spp. is only poorly controlled by seed treatments.

Root rot (*Pythium* spp.) (**Browning root rot**)
Root rots caused by various species of Pythium were first investigated in Canada in the 1930s and the disease they caused was called 'browning' root rot because of the scorched appearance of the young diseased plants. The affected plants occurred in large or small patches and the symptoms included leaf die back, stunting and subsequently poor yields.

In the United Kingdom *Pythium* spp. are common in the spring on the roots of autumn sown wheat, barley and oats. They cause a pale brown, soft rot towards the tips of the young roots but the disease is rarely seen on older parts of the root system. Sometimes root rots are associated with poor growth especially in wet areas and also with a yellowing which may be confused with 'May yellows' (page 175) though the latter affects much larger crop areas. The disease is favoured by cool wet conditions and by soils low in phosphorus. Barley seems to be more affected by the disease than wheat or oats and especially when soils are wet.

The importance of root rots caused by *Pythium* spp. in the United Kingdom is uncertain. They are undoubtedly common on young plants and occasionally can be shown to be damaging. However, it has not yet been possible to assess the damage caused to older plants because of a problem of measuring satisfactorily *Pythium* root rot on the root systems of mature plants. During the past decade in the United States Pacific North-West soil treatments with general soil sterilants and with *Pythium*-specific fungicides have increased yields commonly by 15–25 per cent and these increases have been related to the control of *Pythium* root rot. In the United Kingdom, similar experiments resulted in quite large yield increases from the general soil sterilants but not from chemicals aimed specifically at the control of *Pythium* root rot.

Yields tend to decrease when successive wheat crops are grown, even when known diseases such as take-all are not causing significant damage. The role of pathogens such as *Pythium* spp. in reducing yields in such circumstances has not yet been properly assessed.

Root rot (*Rhizoctonia solani*)
Rhizoctonia attacks a very wide range of plants and is occasionally of significance in cereals. It is damaging most frequently in barley (for fuller account see page 173). It attacks the roots of seedlings and young plants. In severe cases plants are killed, but more usually the infected plants struggle on until the onset of warmer weather in the spring, when secondary root production enables a

degree of recovery to take place. The disease occurs in sharply defined patches, often quite small but occasionally covering considerable areas. Plants are stunted and frequently exhibit a purple discolouration ('purple patch'). At harvest time affected patches are less mature and remain green longer than the healthy crop. Grain yield is negligible.

The disease is not recorded frequently on wheat, largely because damage tends to be associated with light sandy soils on which wheat is rarely grown. The disease is favoured by dry cold springs and is more common in cereals in rotation with roots and grass.

Snow rot (*Typhula incarnata*)
This is uncommon on wheat in Britain. It is noticed when the snow has melted away. The lower leaves become yellow then brown and some plants may be killed, resulting in a thinning of the stand. The base of the plant bears a mould with numerous and characteristic reddish-brown fungal bodies (sclerotia) about the size of clover seed. These sclerotia survive in the soil as a source of infection for the following cereal crop. The disease is more common in Scandinavia and on the Continent where snow persists. In the United Kingdom it is more common in winter barley (see also page 174).

Winter frost damage
In spring, when growth resumes after severe winter frosts, autumn-sown wheat (and other cereals) on soils prone to 'capping' may appear sickly and yellow. Examination of the underside of the soil 'cap' will reveal the severed sub-crown internode. This mechanical injury may be easily confused with soil pest damage. The condition is often referred to as 'frost lift' and results from the heaving of the frozen surface soil 'cap' and consequent snapping of the stem below ground. Rolling to consolidate the soil will often assist recovery of partially damaged plants.

Oats are particularly susceptible to low temperature effects and can suffer badly in prolonged frosts. Aerial parts of the plant become water-soaked; when soil temperatures rise the affected tissues dry out, turn brown and die. In the winter of 1981–2, which was characterised by two separate periods of severe frosts, many crops of autumn-sown oats were lost. In contrast, autumn-sown wheat and barley suffered little or no damage. Frost damage at a later growth is described on page 148.

Take-all (*Gaeumannomyces graminis* var *tritici*, syn. *Ophiobolus graminis var tritici*) (Plates 24–7)
This disease occurs wherever wheat and barley are grown in temperate climates. It is considered to be one of the most important of the diseases of cereals, especially where these are grown intensively. It has been recognised in the United Kingdom for over seventy years. All varieties of wheat and barley are susceptible. Oats can be regarded as practically immune, except in western and northern England, Wales and Scotland, where a specialised strain is capable of attacking all three cereals. Rye is generally less susceptible but occasionally severe attacks have been recorded in rye on land recently limed.

Damage symptoms
The fungus attacks the roots of the plant and is able to do so at any stage of growth. Early infections cause stunting and yellowed foliage and in very severe attacks on early sown autumn cereals, patches showing these symptoms may be seen as early as November or in the spring. When affected plants are dug up, the root system is seen to be blackened in parts, sometimes extensively.

The first symptoms of the disease are most often seen after earing, when crop patches of variable extent appear stunted and some of the worst affected plants are bleached and dead even before flowering It is this stage of the disease which prompts its alternative name 'whiteheads' but bleached, prematurely ripened ears can be caused by other factors such as brown foot rot (page 95), eyespot (page 98), sharp eyespot (page 106) and some pests (page 35).

Following the first appearance of the disease, these patches increase in size more or less progressively until harvest, giving the impression of a rapidly spreading attack. This is not so in reality; the spreading appearance is merely reflecting the later development of symptoms on plants less severely affected. Severe infection results in early whiteheads and 'rat-tail' ears which produce little or no grain. With progressively less severe infections, the damage ranges through slightly fuller ears containing mostly tail corn to ears which, although apparently healthy, fail to 'finish' their grain properly. The 'whiteheads' symptoms may be relatively short-lived in a wet season because the dead ears are rapidly colonised by secondary black moulds (page 145) which turn prematurely ripened patches into black areas long before harvest. Diagnosis of take-all is not always easy, partly because this disease

Plate 24. Take-all in wheat. An aerial photograph using infra-red film showing the association between take-all and soil conditions. The light areas are weeds which have grown in parts of the field severely affected by take-all. The severity of take-all reflects a soil pattern. Gullying in the chalk drift has been infilled with soils of a different texture and structure which are subject to waterlogging. Plants growing in these areas were much more severely damaged by the take-all fungus (ADAS Photographic Unit). Crown copyright.

Plate 25. Take-all. A severe attack on mature wheat plants showing a serious loss of roots and blackening of the stem bases.
National Institute of Agricultural Botany.

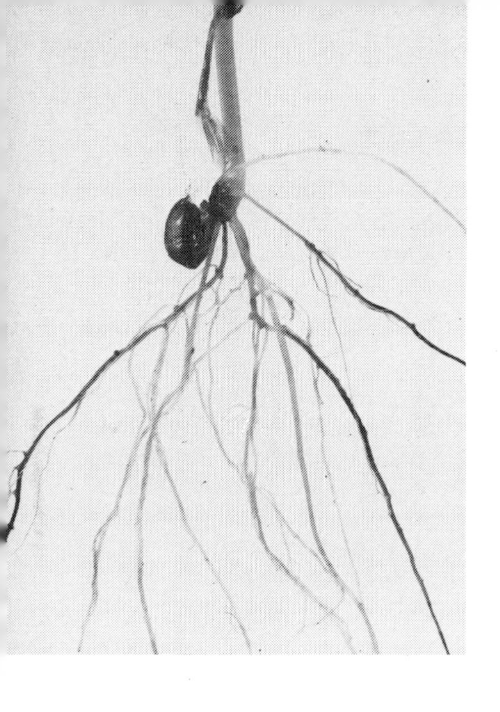

Plate 26. Take-all. A wheat seedling with roots blackened by the disease at an early growth stage.
Rothamsted Experimental Station.

Plate 27. Take-all causing poor stunted growth in a patch in winter wheat.
National Institute of Agricultural Botany.

is often accompanied by eyespot infection on the same plant. However, typical 'whiteheads' accompanied by a black sheath of mycelium exposed when the lowest leaf sheath is removed and (most important) blackened roots are fairly sure indications of take-all. Eyespot (page 98) does not cause a root rot.

Life history

After harvest, the fungus remains alive on the stubble and root residues of the infected crop. It is at this time that the sexual stage of the fungus can be seen as tiny flask-shaped bodies embedded in the stem base tissues, with the neck protruding through the leaf sheath. These spore cases (perithecia) occur most frequently in wet conditions. In dry weather infection is often confined to the roots of the plants, the stem base not becoming black at all. When the stubble is ploughed perithecia may sometimes be found in wet weather on the exposed roots.

The role of the spores (ascospores) produced in these perithecia in the initiation of disease is in some doubt. They are thought to be the means of introducing the fungus into new land areas, such as the polders of Holland, which, by their origin, could not be carrying indigenous infection. In more usual farming situations, it is thought that ascospores make little, if any, contribution to the spread of the disease.

Volunteer plants after harvest are attacked by mycelium surviving on crop debris and can be so badly affected in a warm moist autumn that they are killed. More often, they survive to carry infection through to the following year. Weed grasses can also be infected and the most dangerous carriers of infection are those which produce runners or stolons such as couch, Yorkshire fog and bent grasses.

In the absence of carriers, the take-all fungus is a weak competitor with the masses of other soil fungi and bacteria and it is relatively soon starved. While survival for two years in the absence of a suitable host has been recorded, in practice the take-all fungus is reduced to an acceptably low level in about ten months, which explains why a clean one-year break of a non-cereal crop is generally an effective method of control. Emphasis must, however, be placed on the *clean* one-year break. Volunteers, or couch grass surviving through the break, can often nullify its beneficial effects. Indeed this problem of carriers in the break is responsible for the poorly beneficial effect of the one-year ley in some areas.

Ryegrass leys are very effective breaks. When the grass is undersown, volunteers from the cereal nurse crop cannot be controlled and they may carry the disease. The presence of couch grass in the nurse crop is more dangerous, since it is much more likely to survive the winter and to spread in the ley. For take-all control, therefore, it is essential to plan couch control to ensure that the nurse crop is sown in a clean situation, otherwise the ley break will not successfully control the disease.

Another factor in the success of grass leys as break crops for wheat is the presence on grass roots of a fungus (named *Phialophora radicicola* var. *graminicola*). High populations of this fungus in first wheat soils have been associated with the low populations of the take-all fungus and a consequent reduction in the disease. The mechanism is not understood though it has been referred to as cross protection. The *Phialophora* fungus has also been found in high populations in some soils after crops other than grass. Under such conditions the build up of take-all is delayed for one or two years before the effect is lost.

Many factors influence the spread of take-all in the crop, though this is always slow since the fungus can only grow along the intertwining root systems of adjacent plants. Since the fungus is a root invader, any limitation on root growth or function will make the effects of the disease much worse. Poor drainage and soil compaction are two of the most common adverse factors implicated in severe take-all attacks, and the pattern of disease very often follows closely the pattern of distribution of the poorly structured soil, as shown in Plate 24 (page 88).

Rotations

The fungus is built up to damaging levels by frequent cropping with wheat and/or barley. Wheat is more susceptible than barley and autumn-sown crops are more severely affected than are spring-sown ones. The disease is, therefore, built up most rapidly in successive winter wheat crops and it is in these that damage is greatest. If the field to be cropped has not grown cereals for several years and has been substantially free of weed grasses, two winter wheat crops may be taken with little risk of severe take-all on light land and usually three on heavy land. Barley is less susceptible and can often be grown for several successive years without serious losses if it is adequately manured and particularly if it is spring-sown.

If, however, cereals have figured more recently in the rotation, disease build-up will be more rapid. For instance, the second winter wheat after a one-year break following a run of cereals can be very badly affected. It is preferable, therefore, to restrict the highly susceptible autumn-sown wheat to the first crop after a 'break' of one year and to follow this with the more resistant barley. In more traditional systems, in which cereals may occupy no more than half the acreage, there is little risk of the disease having been built up and therefore the sequence of wheat and barley is less critical.

Continuous cereals and take-all decline

In spite of the foregoing, it is possible to grow even winter wheat consecutively as is illustrated by the famous Broadbalk field at Rothamsted Experimental Station. From experiments there and elsewhere, it has been shown that build-up of take-all occurs under successive winter wheat crops to reach a peak in about the third to the fifth crop, after which, far from increasing as one would expect, the disease actually declines. This phenomenon is known as 'take-all decline'. Later crops show lower levels of disease and a coincident increase in yield though yields do not reach the levels of those in the first and second years. The yield loss in the year of peak disease can be very high (as much as 40 or 50 per cent) and it is largely for this reason that few farmers have put the practice of continuous winter wheat growing to the test. However, some have done so and those on well-structured fertile soils, where the effects of take-all are less severe, have been particularly successful, taking advantage of 'decline'.

Continuous spring barley growing has been more widely practised, especially on the chalk soils in southern and eastern England, and the pattern of disease and yield is very similar. The disease takes longer to increase under barley, reaching its peak in five to six years, before entering its decline phase. The peak disease is often not very severe and the corresponding adverse effects on yield smaller. The decline under barley is therefore not as clear cut as under wheat and its subsequent improvement is also less marked. It seems probable that other, so far unknown, factors are operating in a similar manner to take-all and that they are relatively more important in monoculture of barley than in that of wheat.

Take-all in winter barley has been much less studied but field observations indicate that it probably follows the same pattern as winter wheat in respect of decline but is much less affected by the disease during peak years, in common with spring barley.

The explanation for the decline phenomenon is not known with certainty, but it appears to be microbiological in nature. The phenomenon has been recorded in Europe and North America and appears to be a natural consequence of wheat or barley monoculture. If a crop other than wheat or barley is inserted then the take-all decline is partially and temporarily destroyed.

Manuring

Care should be taken to avoid shortage or imbalance in nutrition. Excessive liming encourages the disease. Adequate nitrogen is essential to counteract loss of root to disease. The fungus survives better when nitrogen is available in the autumn, so for winter wheat and winter barley in risk situations, the crop should ideally be drilled with phosphate and potash only. In the spring such crops should never be short of nitrogen and a split dressing, the first dose applied early, is preferred.

One of the reasons for the rare occurrence nowadays of the classical take-all symptom of dead patches in May is the use of stiff, short-strawed varieties which can utilise high nitrogen applications without serious lodging.

For spring barley, there is no call to depart from the normal practice of applying all the nitrogen to the seedbed or not later than the three-leaf stage.

Other factors

Any shortfall from optimum conditions for root growth will enable take-all to do a disproportionate amount of damage to the crop. Reference has already been made to the interaction between take-all and poor soil conditions. Sowing date also influences the balance between fungus and plant. For winter wheat and winter barley early sowing following a susceptible crop exposes the young root system to higher levels of inoculum. However, sowing after the optimum date will mean fewer well-rooted plants going into winter and the crop will be less able to withstand an attack of take-all. While it is unwise, therefore, to sow very early in high-risk conditions, it is equally unwise to delay drilling beyond the optimum date.

However, the large area now sown to cereals in the autumn and the association of earlier drilling and high yields has forced farmers into some very early drilling dates. In these circumstances the earliest sowing dates should be reserved for fields that are not at risk from take-all.

For spring-sown crops the risk of serious damage from take-all is small. The right course for achieving high yield as well as for disease control is to drill the crop when the seedbed is in good order, irrespective of date.

A well-consolidated soil slows down the spread of take-all along the roots and the effect of this can sometimes be seen as healthy headlands around a diseased field; the opposite picture of a diseased headland surrounding a healthy crop may be an effect of soil compaction or of invasion by infected couch from hedge bottoms and track verges.

All wheat varieties are susceptible to take-all and so far searches for good sources of resistance have proved unsuccessful. Some-wheat × rye crosses (triticale) have shown some resistance and could provide a useful means of some control on land which is not considered the best for wheat growing and where take-all damage can be severe.

Control

In the absence of resistant varieties a complete control can only be obtained by adequate rotations. However, with such a high proportion of the arable area in cereals this often does not provide a practicable solution and in these circumstances the effects of take-all can be reduced by good husbandry, especially cultivations designed to encourage maximum root growth, optimum nitrogen use including some applications in early spring where necessary and the avoidance of very early sowing where there is a risk of severe disease. Well structured, fertile soils, well managed, can sustain intensive, even continuous winter wheat at economic levels of yield. However, it is not possible to predict severe take-all attacks and farmers have to accumulate experience to identify the soil and conditions which will permit this kind of cropping. On less favourable soils, winter wheat is best restricted to the first crop after a break or alternated with non-susceptible crops. On such soils triticale wheats may prove useful.

There are no effective chemical control methods but at least one seed treatment (triadimenol) can reduce the severity of the disease. The control may not be effective when disease potential is very high but reductions in disease, especially in some early drilled crops have been demonstrated and may persist into the spring. Such control is sometimes, but not always, associated with yield increases.

Recently attempts have been made to utilise 'biological control' of take-all. Some soil-borne organisms are known to be antagonistic to the take-all fungus and probably play a part in the take-all decline phenomenon. Some bacteria (mainly *Pseudomonas* spp.) from soils exhibiting take-all decline have been isolated and multiplied and then applied to seed before sowing. In some experimental plots, especially in North America, a good control of take-all has been obtained but in general the results have been inconsistent and are not yet at a stage where they can be applied to commercial crops.

Seedling blight, brown foot rot and ear blight (*Fusarium* spp.) (Plate 28)

These diseases are caused by a group of seed-borne and soil-borne fungi belonging to the genus *Fusarium**—*F. culmorum*, *F. nivale*, *F. graminearum* (see also under scab, page 144) and to a lesser extent *F. avenaceum*. *F. poae* is sometimes recorded on ears. The diseases caused by these fungi are common though they rarely cause severe losses in individual crops. They fall into three distinct phases—a seedling blight, a brown foot rot often resulting in 'whiteheads' and an ear blight.

Seedling blight

Infected seedlings may be killed before or soon after emergence. Those which survive have brown marks on the stem bases and the root system is affected by a brown root rot. Severe attacks occur in dry soils and, whilst *F. nivale* is favoured by low soil temperatures, the other species are favoured by high temperatures. Severe seedling losses are sometimes associated with the use of seed not treated with an effective fungicide. *Fusarium* spp. are also fairly common soil fungi, persisting as spores or in dead plant tissues. They affect a wide range of crops and other plants in addition to cereals. The relative importance of seed and soil as sources of the fungi is not known.

Under most farming conditions seedling losses are negligible and are not noticed. Slight losses are compensated for by increased tillering of healthy plants. In a few cases a thinned stand can result in yield reductions although these are rarely serious. In winter cereals under snow, *F. nivale* can cause a disease known as 'snow mould' (as distinct from snow rot, page 86) in which leaves and sometimes entire plants are yellowed and may be killed. A characteristic pink mould may be present on the dead tissues. This disease is sometimes present with snow rot especially on winter barley (page 174).

* The other names for these fungi (perfect stages) are *F. nivale* = *Calonectria nivalis* (syn. *Griphosphaeria nivalis*): *F. avenaceum* = *Gibberella avenaceae*; *F. graminearum* = *Gibberella zeae*.

Brown foot rot and 'whiteheads'

These are probably the most common symptoms caused by *Fusaria*, mainly of the *F. roseum* group such as *F. culmorum*. Occasionally the disease may be noticed in wet autumns and more so when the weather gets warmer in the spring. Patches of diseased plants then become obvious, in comparison with healthy plants, by their poor growth and yellow leaves, which frequently show a tip-scorch. If plants are examined at this stage, the rot will be seen to have involved not only substantial proportions of the root system, but also the stem base, which presents a brownish,

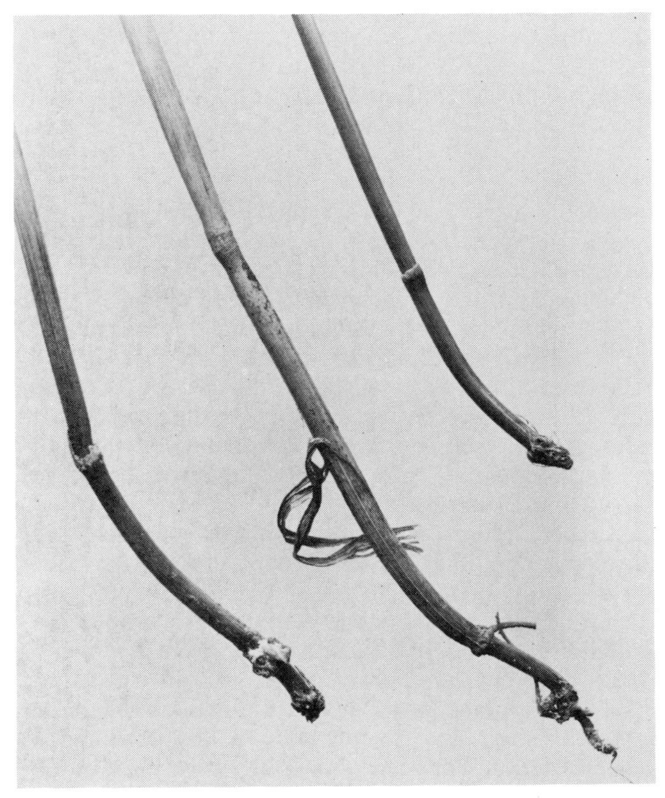

Plate 28. Brown foot rot (*Fusarium* spp.) of wheat. A severe rot (centre tiller) and a stem rot causing the tillers to break off just above soil level (outer tillers). The pink-white *Fusarium* mould is on the lower joints (nodes). National Institute of Agricultural Botany.

somewhat water-soaked appearance. Plants invaded to this extent often fail to ear.

A more usual symptom (though still relatively uncommon) mainly affects the stem with much less root rotting evident, and leads to premature death of the plant at any time between ear emergence and harvest; it also causes 'whiteheads' and ears with shrivelled grain. The earlier the death, the more shrivelled the grain. In the field, plants of this kind occur scattered through the crop and are usually easily knocked over or pulled up. When the base is examined, apart from the brown wettish discolouration it will be seen to have broken off at one of the lowermost joints (nodes), leaving the root system behind in the soil. The fungus responsible for the attack can be seen as pinkish white, pinhead-sized pustules on the lower 'joints' of infected plants. In north western United States of America, symptoms similar to these are associated with serious crop losses but occur only when the plants are subjected to severe water stress.

A form of brown foot rot is also often noticed in crops in the United Kingdom that have been under stress, especially those affected by drought. This usually becomes obvious after earing and can be much more extensive than in the cases described above, often following well-defined soil patterns where the effects of drought are worst. In these cases, however, the stem bases, although brown and infected by *Fusarium* spp., are rarely rotten and all plants are similarly affected, not a mixture of diseased and healthy plants. In attacks such as these the fungus is usually regarded as a secondary invader, with drought as the real cause of the damage.

In wet conditions, the dead 'whiteheads' are rapidly colonised by moulds and turn black (see page 145).

Ear blight and leaf disease

The diseased stem bases bear the characteristic sickle-shaped spores of the fungus. These are blown and splashed in moist weather and infect the ears to cause the ear-blight phase of the disease. Ears of otherwise healthy as well as of diseased plants may be infected. Infection may be limited to individual spikelets but when the rachis (stem) of the ear is affected, all parts of the ear above the diseased part of the rachis become conspicuously bleached and grain fails to fill. The fungi mainly involved are *F. culmorum* and *F. avenaceum*. Occasionally the ears may be partially or completely covered by pink-red spore masses. This phase has been called 'red mould' (see also scab, page 144). The grain in affected ears can become infected at the same time and spores of the fungus can also contaminate healthy grain. In these ways, the disease can be seed-borne and, although seed treatment with fungicides helps to reduce seed-borne contamination, it cannot eradicate true seed infection.

In another phase of the disease caused by *F. nivale*, ascospores produced from the numerous black perithecia on the sheaths of stem bases, become airborne and are carried to upper leaves and ears. On the leaves infections show as large pale brown spots and on ears as pale spots with dark brown margins on the glumes. Both symptoms are relatively uncommon in the United Kingdom, but in France the ear symptom is considered important and fungicide sprays are applied immediately after ear emergence to control it. In the United Kingdom *F. poae* has also been associated with distinctive spots on the glumes.

Control

The fungi concerned are normal inhabitants of agricultural soils and in the United Kingdom rotation is not known to have any effect on the disease. However, in some other countries crops such as oats and maize are known to favour the build up of the fungi, especially *F. culmorum*, in the soils. Although the seed-borne phase of the disease can be only partially controlled by the use of organomercury and other seed treatments, it is important that they be carried out because occasionally losses from seed-borne infection can be very serious. There is evidence that such treatments also give some protection against the soil-borne phase of *Fusarium*. Some varieties of wheat are more susceptible than others to *Fusarium* diseases, especially the ear blights and these diseases are now receiving attention from plant breeders.

Eyespot (*Pseudocercosporella herpotrichoides*, syn. *Cercosporella herpotrichoides*) (Plates 29–30)

This disease, which affects wheat, barley and (less frequently) oats and rye is regarded as a soil-borne disease because it survives on infected stubble. Nevertheless, it spreads in the crop mainly by rain-splashed spores and is much influenced by weather conditions, especially temperature, as well as by rotation. The fungus spreads most in the wetter, cooler months of the year and is mainly a serious disease of autumn sown crops though occasionally, and especially in the wetter western areas, it can be damaging in spring sown crops.

Plate 29. Eyespot in winter wheat causing lodging with the straws falling in different directions.
Rothamsted Experimental Station.

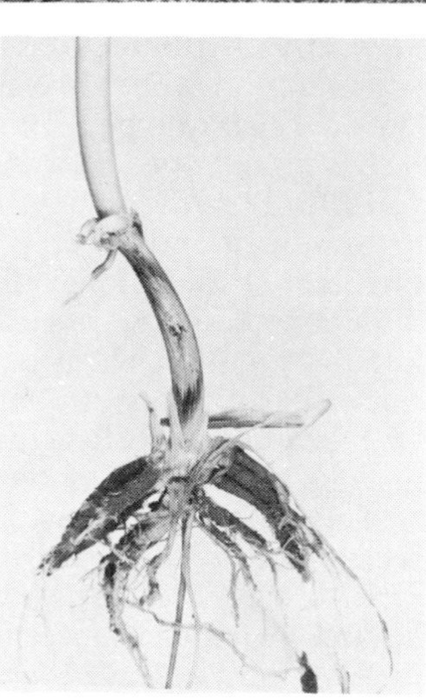

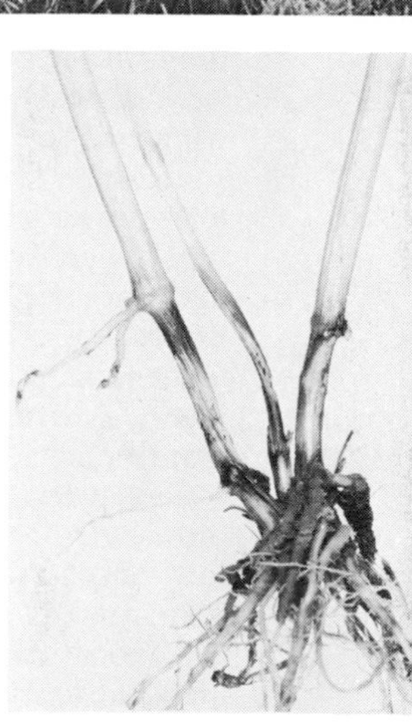

Plate 30. Eyespot in wheat. (Left) Lesion on a young plant; (centre) lesion on a mature stem; and (right) severe damage with stems twisting at the lesions.
Rothamsted Experimental Station.

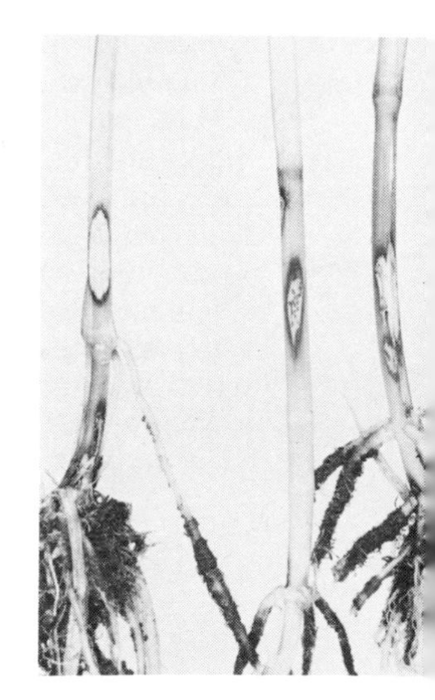

Plate 31. Sharp eyespot showing the lesions with sharply defined margins and with dark fungal cushions on the lesion (middle stem).
Rothamsted Experimental Station.

The fungus
The fungus which causes eyespot exists in two forms: the W-type isolates which attack wheat more than rye and the R-type isolates which attack wheat and rye equally. The two types can also be distinguished by their growth in agar culture – the fast, even growth of the W-type and the slow, feathery growth of the R-type*.

In the 1970s the W-type isolates were associated with eyespot in wheat and the R-types were readily obtained only from areas where rye was grown extensively. In 1981 the R-type isolates were found to be more widely distributed and in a survey in 1983 they were at least as common on wheat and much more common on winter barley than were the W-types. This shift in the population coincided with the development of resistance in the eyespot fungus to the widely used MBC fungicides. A higher proportion of R-types than W-types were found to be resistant to MBC and it therefore seems likely that these fungicides selected in favour of the R-types. In addition, R-types are relatively more pathogenic to winter barley than to winter wheat so that the large increase in the area of winter barley since the late 1970s may also have contributed to the selection of R-types. There is also some evidence that the widespread use of the 'DMI' fungicides (see page 26), to which the R-type isolates are less sensitive, meant they may have selected in favour of the R-types.

In respect of the disease they cause and its symptoms, their effect on varieties and their persistance in the field the two types are believed to behave similarly.

Symptoms
Infection can take place soon after seedling emergence and in severe attacks, such as can occur in early autumn sowings, the young seedlings can be killed outright. More typically the symptoms are seen in spring. In the early stages of attack the outer leaf sheath, just above soil level, bears a rather diffuse brown smudgy lesion on one side, which at this stage has no effect on the plant. As the infection progresses, penetrating deeper into the outer leaf sheath and subsequently into the tissues below, the lesion becomes more defined, developing into an eye-shape with a somewhat indefinite outline and a 'pupil' of black dots.

Depending on the stage of development at which infection takes

* The W-type isolates may be referred to as *P. herpotrichoides* var. *herpotrichoides* and the R-type isolates as *P. herpotrichoides* var. *acuformis*.

place and on the rate of penetration of the tissues, individual infected shoots, especially small secondary tillers may be killed in the early stages before stem extension occurs. More typically, however, the disease develops relatively slowly, eventually penetrating the stem and, after stem extension, producing typical eyespot lesions on the straws, usually near soil level. The severity of the symptom and the damage caused depends mainly on the susceptibility of the variety and the effects of the environment on lesion development. The straw may be so weakened by the attack that it twists in the middle of the 'eye' and falls over. In the more resistant varieties the stem lodges less readily but severe lesions cause whiteheads. When infected straws are split open the hollow stem in the region of the eyespot is often, though not always, filled with a grey fungal growth. Symptoms of eyespot are, at some stages, similar to those of sharp eyespot (see p. 106).

Before the more resistant varieties were introduced the most severely attacked stems of susceptible varieties would topple and 'straggle' among the upright stems throughout the period from ear emergence to harvest. When a high incidence of severe attacks occurred, sufficient straws would be damaged to carry the whole crop down, with straws of the laid crop lying in all directions. Modern varieties are not only more resistant to eyespot but are also stiffer strawed so that straggling is not a common symptom. Some severely affected tillers remain standing but die prematurely and appear as 'white heads' containing shrivelled grain. The white heads are usually colonised by secondary moulds and turn black (see page 145). Occasionally the attacks can be severe enough to cause lodging even in the modern resistant varieties.

It is as a cause of severe lodging that eyespot is most damaging, although of course not all lodging is caused by eyespot. Over-generous nitrogen manuring, particularly if given in the early spring, can lead to weak straw and lodging, especially if the crop is a potentially heavy one. In these circumstances the lodging is usually observed in well-defined patterns, often in swaths across the field related to the path of the fertiliser distributor or to a soil pattern, with the straws most often lying one way dictated by wind direction.

Two particular aspects of lodging serve to distinguish that caused by eyespot from 'non-parasitic' (nitrogen or weather) lodging. Where eyespot is the cause, examination of straw bases will show a large majority with eyespot lesions and the straw is broken or characteristically twisted in the middle of the 'eye'. Furthermore, the lodged straws lie flat and close to the ground and remain there. In 'non-parasitic' lodging, the straw does not break but bends over at the base and the stems just below the ears may sometimes assume a vertical position by reason of differential growth at the upper joints, these behaving rather like the elbow joint. This partial recovery to an erect position does not occur with eyespot lodging because by this stage the base of the straw is so completely rotted that no further stem growth can occur.

Survival and spread

The fungus survives on the stubble remaining from an infected crop. That which lies on the surface after ploughing and seedbed preparation produces abundant spores on the old lesions during damp weather in the autumn and these are carried in rain splashes to the newly emerged crop. Further spread can occur from spores produced on straws and on infected plants throughout the winter and in fact throughout the life of the crop when favourable environmental conditions occur, though usually spread is greatest before the onset of drier, warmer weather. Most abundant spore production occurs at relatively low fluctuating temperatures, with an optimum at 5–10°C and decreasing as temperatures increase, up to 25°C.

The fungus can survive on buried straws for two or three years and at low levels even longer. When stubble residues from an infected crop are brought back to the surface by ploughing, they are still able to produce spores freely and these can start an epidemic. For this reason a one-year break, which is usually effective for take-all control, is quite ineffective for the control of eyespot.

Apart from carry-over on infected stubble debris, self-sown wheat and barley may serve to carry infection from one crop to the next. The role of grasses in the survival of eyespot is not certain, although many grass species, especially couch (*Agropyron*), can be infected by the fungus. In practice, however, the importance of these sources of infection is small compared with infected stubble and a grass break of two or more years' duration provides a very effective method of control.

Disease Development

Infection is favoured by wet conditions and cool temperatures, optimum about 9°C. The incubation period before symptoms are seen is 4–12 weeks, the shorter period being associated with higher temperatures. The subsequent development of lesions is not fully understood but appears to be a result of several interacting factors.

The lesions are first seen on the outer leaf sheaths. The subsequent penetration of the inner leaf sheaths and development of lesions on the stem are favoured by above average temperatures. However, such temperatures are also associated with dry weather which increases the rate of death and shrivelling of the infected leaf sheaths and thus hinders the development of stem lesions. It is not surprising therefore to find that eyespot development is favoured by wet weather and the humid conditions provided by wet soils and dense crops.

Effect on yield

For assessing eyespot severity and measuring effects on yield, infections are classified as 'slight', 'moderate' or 'severe'. These are defined as:

slight lesions affecting less than half of the circumference of the stem;
moderate lesions affecting half or more of the circumference;
severe as 'moderate' but with the stem so weakened that lodging could occur.

Most yield loss is attributed to the 'severe' category, where on average about one-third of the potential yield is lost; straws in the 'moderate' category may lose up to one-tenth of potential yield, while 'slight' infection has no measurable effect. In surveys during 1975–80 in England and Wales, these assessment methods and yield loss estimates indicated an annual loss from eyespot of about one per cent of the England and Wales wheat yield. This relatively small loss, when a significant proportion of crops were at risk to the disease, reflects the success of breeding for resistance to eyespot. During the 1980s there was evidence of increased losses in some years, probably associated with changed cultural practices which favour the disease, such as early sowing and the growing of denser crops. Furthermore, field observations suggest that the 'moderate' category of damage was sometimes associated with more severe losses than those indicated by the earlier disease assessments.

The main components of yield affected by eyespot are numbers of grains per ear and average grain weight, and the adverse effects are always greater if lodging occurs following severe infection.

Control

Eyespot can normally be kept under control by a rotation which provides at least a two years' break from wheat or barley. This is usually sufficient to reduce the residual fungus to a level that is unlikely to initiate a significant attack. However, occasionally severe attacks have occurred after a two-year break and these are probably due to the exposure by ploughing of some deeply buried straw residues which have remained undisturbed by cultivations during the intervening years. Oats and leys are generally as effective as non-cereal breaks in practice even though they too can become infected.

Because of the intensity of cereal growing in the United Kingdom, control by rotation is often not possible so the main method of control is by the use of resistant varieties. Most of the wheat varieties commonly grown exhibit an effective level of resistance to the disease (see NIAB Recommended Varieties). Varieties with the highest level of resistance should be grown where there is a risk of eyespot attack; varieties with lower levels of resistance are best grown after a two-year break from wheat or barley. Eyespot lesions occur just as frequently in resistant as in susceptible varieties but the rate of penetration through the stem tissues is reduced. This, and the inherently stiffer and shorter straw of modern varieties, substantially reduces the risk of severe losses directly and through lodging. Nevertheless, under conditions which are favourable for the disease, severe attacks can cause losses even with resistant varieties. The resistance to eyespot used in breeding was first discovered in the variety Cappelle Desprez and has remained effective for a long period. A new and better source of resistance (in the line VPMI derived from *Aegilops ventricosa*) is now available in new varieties, in addition to the Cappelle Desprez source, but it is not known if it will prove to be as durable.

Stubble cultivations appear to have little effect on levels of infection and, although stubble burning with the aid of oil or gas burners has been claimed to be effective on the Continent of Europe, experiments in Britain with this type of machine have not been successful in reducing disease.

Very early sowings in autumn favour severe attacks because the plants are exposed longer to infection and because such crops tend to produce a dense growth (high tiller population) which favours disease development. Other factors which favour dense growth such as high seed rates and high levels of nitrogen fertiliser (especially when early applications cause increased tillering) also favour disease development as will wet weather.

Control with chemicals

The use of straw-shortening chemicals such as chlormequat enables crops to resist lodging but has no direct effect on the damage caused by eyespot.

MBC-fungicides (containing benomyl, carbendazim or thiophanate-methyl) were the first fungicides to be used for the control of eyespot in the United Kingdom, from about 1974 and following previous experience with them in Europe and North America. Sprays applied at about the first node growth stage (GS31) gave a variable control, sometimes up to 80 per cent but on average only about 40 per cent. Although serious attacks of eyespot were not common, the chemicals became widely used because they were cheap, provided control where necessary and gave a cost effective return, increasing yields by about 2 per cent even when eyespot was not damaging. By 1982 about half of the crops of winter wheat and winter barley were treated. In 1981 a few cases of failure to control eyespot in winter wheat were attributed to the emergence of strains of the eyespot fungus resistant to the MBC fungicides. Surveys in 1983 and 1984 showed that such resistance was present in more than half of the crops so that the fungicides could no longer be safely recommended for the control of the disease. The majority of resistant isolates were of the R-type and as mentioned previously the selection pressure exerted by the widespread use of the MBC fungicides was probably the main reason for the emergence of the R-type isolates as the most commonly found form of the eyespot fungus.

The only alternative to the MBC fungicides at present is prochloraz which provides control of both MBC-resistant and sensitive strains, though in earlier tests it proved to be inferior to the MBC fungicides for the control of the sensitive strains.

As in the case of MBC fungicides, the application of sprays containing prochloraz have usually given cost effective yield responses. In fact, in recent years the mean responses of 5–7 per cent were higher than those associated with the MBC fungicides. Such increases were sometimes but not always associated with the control of eyespot. The reason for yield increases when eyespot control is not a significant factor is not known but may be associated with the control of other diseases such as those caused by *Septoria* spp. At high-eyespot sites fungicide use can result in much higher yield increases.

Up to the present it has proved difficult to provide adequate criteria on which to base decisions on whether to spray and how to time sprays. This is because it has not proved possible to predict the occurrence of damaging attacks of eyespot. Even when the disease is common at the first node stage, or if weather favourable to infection occurs (as proposed in a scheme from West Germany), damaging attacks do not necessarily follow if conditions do not favour lesion development. Two kinds of criteria have been proposed tentatively, based on crops at risk and on the presence of eyespot symptoms. Firstly, crops at risk are those where there is less than a two-year break, sown early (e.g. September), where the growth is dense and where the resistance of the variety is less than the best. Secondly, crops are examined for eyespot symptoms at the stem erect stage (GS30) to the second node stage (GS32) and crops with at least 20 per cent tillers affected are deemed to be at risk. On average the best spray timing is at first node (GS31) but earlier or later sprays may be advised depending on the development and the severity of the eyespot symptoms.

Sharp eyespot (*Rhizoctonia cerealis*) (Plate 31, page 99)
The fungus causing sharp eyespot is now recognised as *Rhizoctonia cerealis* which is distinct from *Rhizoctonia solani* formerly, and somewhat tentatively, identified as the cause. *R. solani* is associated with a root rot (page 85) while *R. cerealis* does not affect the roots of cereals.

The disease affects wheat, barley, oats and rye. The pathogen exists in a number of strains which differ in their virulence to particular hosts. At the seedling stage, at least, barley is the least susceptible cereal, with wheat less susceptible than oats, and rye the most susceptible.

In wheat the disease is common and widely distributed though it is only occasionally severe enough to cause significant yield losses. Severe damage was reported from crops in some years in the early 1960s. Then followed a period when the disease attracted little attention and it was assumed to be of no importance. However in the 1980s the disease became much more prominent and in some years the incidence and severity exceeded that of eyespot.

Symptoms
R. cerealis can cause pre-emergence and post-emergence damping-off of seedlings. Sometimes a rot of part of the coleoptile causes an aperture through which the shoot may emerge as a loop giving rise to peculiar deformities. The fungus can also cause the death of established seedlings. All these early phases of attack can result

in a thinning out of the crop, though usually with little effect on yield, and in fact are rarely noticed in the field.

At stem extension to first node growth stages the lesions caused by the fungus may be confused with those of eyespot (page 100) and it is important to distinguish between them since the latter disease is controlled by fungicides that do not control sharp eyespot. At this stage the lesions occur on the outer leaf sheaths, often remaining superficial but sometimes penetrating more deeply. The lesions, mainly eye-shaped and about 1cm long, are more sharply defined at the margins than are the lesions of eyespot. The area inside the brown-purple border is pale cream and often rots away leaving a characteristic shredded area. However, a microscopic examination for the characteristic fungal structures is often required to confirm diagnosis of the disease at this stage.

Later, after stem elongation, lesions occur on the stem and are usually quite distinct from those of eyespot. They may be oval but are generally angular and asymetric. The area inside the narrow brown-purple border is cream coloured and towards the centre the loose cushions of brown fungal mycelium can be easily scraped away with a finger nail. These sharp eyespot lesions are usually smaller, 1–3 cm, often more numerous and can occur higher up the stem (up to 20 cm above the soil) than eyespot lesions. They usually do not penetrate deeply and therefore are not associated with significant damage in terms of yield loss. With this type of lesion there is no grey mould in the straw cavity near the lesion, again in contrast to eyespot.

Sometimes after the completion of stem elongation the lesions become very large and the symptoms are then quite distinctive. It is this form of the disease that has made it so prominent in the 1980s. The lesions on the sheaths, often well above the stem base and 5–10 cm long, may occupy the space between more than two nodes, and they have conspicuous dark brown-purple borders. The white mould of the fungus, sometimes with large brown or black flat sclerotia, is very obvious between the sheath and the stem. The fungus does not always penetrate the stem but when it does it can cause extensive rotting of the stem tissues. With this type of lesion the straw cavity is usually filled with white fungal mould and occasionally may also contain large black sclerotia.

Cases of straw lodging caused by sharp eyespot can occur and are usually distinguished from other kinds of lodging by the fact that the stem is made very brittle and the straw falls over at a clean break of the stem between joints. Occasionally sharp eyespot is the cause of extensive crop lodging. Where lesions are severe but lodging does not occur, translocation of nutrients taken up by the roots is prevented with the consequent death of the plant and the appearance of randomly scattered 'whiteheads' in the crop. These are soon colonised and discoloured by secondary black moulds (page 145), especially in wet weather.

Effects of the environment

Work in the 1960s indicated that the disease was more likely to occur on light sandy soils with a neutral or slightly acid reaction. Cool dry soil conditions favoured infection of the seedlings but the influence of temperature was relatively small compared with that of soil moisture. The disease tended to be worse in winter crops sown later in dry autumns or when soils were dry in the spring.

More recent observations have shown sharp eyespot to occur in crops on all soil types including the heavier soils on which wheat is often grown. The effect of sowing date has not been very significant though late sowings have been less affected. Unlike eyespot, sharp eyespot is not a disease restricted to intensive cereal growing. It can occur under such conditions but severe attacks are just as frequent in cereals grown in a rotation with other arable crops and grass. The saprophytic survival of the fungus on straw or other plant material or as fungal resting bodies in the soil is probably a major factor in the persistence of the fungus between cereal crops grown in succession. However, the fungus can also survive parasitically on a range of crops including grass, peas, potatoes and sugar beet without causing obvious disease symptoms.

Crop losses

Damage to seedlings rarely has an appreciable effect on the yield of winter wheat because the crop is able to compensate very efficiently for a reduction in stand. There is no clear picture of the losses caused by later attacks mainly because effective fungicides have not been available to make comparisons of crops with and without the disease. A few cases of severe losses associated with lodging or whiteheads are recorded in most seasons. However, sometimes losses are apparently less than would be expected, possibly because the severe lesions develop too late in the season to seriously affect grain yield. Early severe lesions may result in complete loss of grain on infected tillers.

A disease assessment method using single tillers has indicated that losses of grain per ear are similar to those caused by eyespot (page 98). Losses associated with severe lesions (girdling at least half the stem and causing stem weakening) range up to 40 per cent with an average of 26 per cent. Losses associated with moderate infections (girdling at least half the stem) were about 5 per cent whilst slight infections (girdling less than half the stem) did not affect yield. Using these assessments in conjunction with national surveys estimated national annual losses over the past decade ranged from less than 0.1 to 1.0 per cent.

Control

There are no effective control measures.

Varietal differences in the susceptibility of wheat do occur but most varieties can be damaged under conditions favouring the disease. Up to the present no fungicides have provided a consistent and effective control. In any case the disease occurs too erratically to recommend routine use of fungicides and at present there are no adequate criteria to select crops which might benefit from fungicide treatment.

Foot rot (*Cochliobolus sativus* [*Drechslera sorokiniana*, syn. *Helminthosporium sativum*]).

This disease is widespread and important in warmer cereal-growing countries but in Britain it has been recorded only a few times on wheat, barley (page 177) and rye.

The fungus which is both seed- and soil-borne infects the roots and may cause the death of seedlings but normally infected plants grow to maturity, the disease causing brown spots on the lower leaves and a rot at the stem base resulting in poorly filled ears.

Seed-borne infection is partially controlled by disinfection with organomercury seed treatment (page 23) and the soil-borne phase by crop rotation.

Yellow rust (*Puccinia striiformis*) (Plate 32)

Yellow rust, sometimes called 'stripe rust', is a serious disease of wheat in Britain and also affects barley, rye and many grasses but not oats. It is an obligate parasite, growing and surviving on living green plants. Many forms of the fungus exist and the form on wheat normally affects wheat only and the forms on other hosts do not affect wheat. Thus grasses are not important as a source of wheat yellow rust in Europe though they are in some other parts of the world. Within the wheat form many specialised races

Plate 32. Yellow rust in wheat. An aerial photograph using infra-red film in which the dark spots indicate small patches (foci) of yellow rust with a halo around them representing secondary infection. The field with the larger number of foci was drilled two weeks earlier than the adjacent and less affected field. (ADAS Photographic Unit).

exist and these are distinguished by their reaction on a range of wheat varieties.

Symptoms

Outbreaks of yellow rust are often first noticed in May as small yellow patches (foci) scattered at random in the crop. The disease may then spread throughout the crop and further afield unless the weather is unfavourable, when the rust remains confined to the patches.

The characteristic symptoms of yellow rust are seen on the leaves as orange-yellow pustules (containing uredospores) arranged in lines. There may be a few or several such lines parallel to each other usually producing a stripe of affected tissue. Sometimes the affected tissue may occur in patches, especially on leaves of young plants, but the characteristic arrangement of the pustules in lines is usually also present. Under conditions not favourable to the disease, affected tissue is pale brown and pustules can only be seen with difficulty. Yellow rust also affects the leaf sheath and the ears. Later in the season the fungus produces black pustules containing teliospores. The teliospores produce another kind of spore (basidiospores) but these appear to be functionless since they are not known to infect wheat or any other host. Thus yellow rust has no known alternate host and relies on the yellow uredospores as the only means of spread.

Sources and disease development

Yellow rust survives the harvest period on late green tillers and then passes to volunteers and finally to winter wheat. As yet there is no evidence that spores survive on or in the soil to provide a source of infection for volunteers or a very early sown crop (as in the case of brown rust) though this might explain cases of early infections where there is no evidence of late tillers. The fungus may produce a few pustules on winter wheat during the winter but it also survives as mycelium within the leaf tissue and does not become apparent until pustules are formed when the weather is more favourable in the spring. Low winter temperatures may kill pustules but mycelium within the leaves will survive to −5°C unless the lower leaves, which are the ones most likely to be infected, are themselves killed by low temperatures.

The optimum temperature for disease development is relatively low, about 10–15°C, so that the disease may spread as early as April and May. Free water is necessary for infection to occur. Subsequent development of the disease is largely dependent on temperature. If the summer is cool and wet the disease will continue to spread but higher temperatures restrict or inhibit disease development and a prolonged spell of warm dry weather (e.g. more than 20°C) will check the epidemic. These effects of higher temperatures are related to increasing host resistance as well as to a direct effect on the fungus. A prolonged period of high temperatures in the summer and at harvest will have such an adverse effect on the disease that very low levels of fungal inoculum will survive to infect the winter crop and this in turn will lead to a late or very slow development of the epidemic in the following spring, even if weather conditions are favourable. High summer temperatures and associated low rainfall not only restrict disease development but also reduce the numbers of late tillers and volunteers, the only means by which the fungus can survive the harvest and post-harvest period, thus decreasing the inoculum available to the following crops.

Epidemics

Epidemics of yellow rust are likely to occur when a significant proportion of the wheat area is occupied by a susceptible variety (or varieties) and the weather conditions are favourable. Since most of the varieties selected by farmers are resistant, such occasions will occur when the resistance of a popular variety breaks down because of infection by new races of the fungus. There is usually a period of more than one year between the establishment of new races and the occurrence of a general epidemic. Thus the early development of an epidemic in any one year depends on susceptible varieties becoming infected in the previous year; on favourable conditions (mainly wet weather) for the survival of the fungus on late tillers and volunteers; and on a mild winter. The epidemic will then develop according to the summer temperatures and rainfall.

Yellow rust spreads through airborne uredospores. Local sources of infection are very important but long-distance spread of the fungus can also occur. Transport of spores in air currents over several hundreds of miles has been demonstrated in North America and India.

Effects on yield

Yield losses caused by severe epidemics in individual crops may be as high as 50 per cent but in the United Kingdom the average loss in an affected crop in an epidemic year is in the range 5–20 per cent. Severe losses are associated with epidemics which start early and continue through the summer with proportionately lower losses occuring when the epidemic starts late or is curtailed by high temperatures. From data collected in surveys in England and Wales, 1967–80, it was estimated that the average annual loss in yield from yellow rust was about 0·2 per cent (range 0–1·2 per cent). In view of the high potential losses from yellow rust these relatively small losses reflect the success of the policy of excluding highly susceptible varieties from the list of recommended varieties.

Since 1980 losses from yellow rust have been negligible. This is probably due to the continuing recommendations of resistant varieties. It is also due to the widespread use of broad spectrum fungicides, especially in the period after the flag leaf has emerged. These are normally aimed at other diseases but are also effective against yellow rust.

Yellow rust affects the plant in at least three ways: it reduces the rate of photosynthesis, increases the rate of respiration and reduces the rate of flow of food materials from the leaves. It affects yield by reducing both grain numbers and grain weight, and infected plants produce samples of shrivelled grain. Root growth can also be severely affected so that plants are more susceptible to drought. Because of this and the development of the resistant reactions in hot dry weather, leaves of rusted plants become desiccated prematurely and this results in the most serious yield losses.

Control

Yellow rust is best controlled by the judicious selection of wheat varieties. In particular it is very undesirable that a large area of wheat on any one farm or district should be at risk to attack by the disease by growing a limited range of varieties known to be very susceptible or containing the same specific resistance. The best solution is to diversify in selecting varieties, and to use mainly varieties known to have good general resistance. When varieties known to contain specific resistances are grown, a selection should be made such that each variety carries a different source of resistance. A varietal diversification scheme for reducing the spread of yellow rust, in which varieties are grouped according to the resistances they contain, is made available each year in the United Kingdom (page 22).

Because the disease spreads rapidly by wind-borne spores, good husbandry alone is unlikely to lead to satisfactory control. However, measures which reduce the chance of autumn infection, which in turn leads to early establishment of the disease in the spring, can have a significant effect on the severity of the disease. Late tillers and volunteers should be destroyed before winter wheat plants emerge and this is more effective when achieved over a substantial area. Crops sown very early are much more likely to be infected in the autumn than those sown later, but in deciding sowing dates the optimum time for crop yields should be the decisive factor rather than the avoidance of yellow rust. Nevertheless, the very early sowing of the more susceptible varieties should be avoided. Growing the same variety in the same field in successive seasons also increases the chance of disease carry-over with the same specific virulent strain. Where this practice is necessary, e.g. for seed production, then the more susceptible varieties should not be sown very early and special efforts should be made to control volunteers.

Fungicides are available for the control of yellow rust and as mentioned above, the relatively infrequent attacks of yellow rust in recent years is probably at least partly due to the widespread use of fungicides which are aimed at other diseases but are also effective against yellow rust.

With a wise choice of varieties the disease will not occur frequently enough to make routine prophylactic application of fungicides economically worthwhile. Where the disease does occur it is suggested that a fungicide should be applied as soon as it is noticed in the very susceptible varieties or when it is slight but generally distributed in moderately resistant varieties. If this is early, further applications may be needed. The more resistant varieties are not likely to benefit from fungicide use.

Brown rust (*Puccinia recondita*, syn. *P. triticina*)
Brown rust, known as leaf rust in North America, is a common disease of wheat and occasionally causes significant yield losses in Britain. It became more common and damaging with the introduction in the mid-1970s of some susceptible varieties. The disease is also favoured by early autumn drilling. This has become more common and increases the risk of autumn infection and thus higher disease levels in the spring which can lead to earlier epidemic development.

P. recondita is an obligate parasite which grows only on living green plants. It affects wheat, rye and some grasses but specialised forms exist so that the form on a particular crop or grass is usually confined to that crop. This fungus is known to have several alternative hosts (species of *Thalictrum*, *Anchusa* and *Isopyrum*) on which the cluster-cup (aecidial) stage occurs but this has not been found in Britain.

The brown pustules containing uredospores occur mainly on the leaf blades though they may also be found on the sheaths and ears. The pustules have a scattered distribution and this, together with their colour, enables brown rust to be distinguished from yellow rust. When significant attacks occur on the flag leaf and the leaf below the flag, the pustules are then often grouped in patches on the lower part of the leaf or on the 'bend' of the leaf where surface wetness from dew or rain tends to persist. Black pustules containing teliospores are formed late in the season but in the absence of an alternative host these have no function. Recently in North America it has been shown that spores can survive in or on the soil and provide a source of infection for emerging wheat seedlings, usually as volunteers. The disease spreads from late tillers and/or volunteers to the winter wheat and thus survives the winter. The disease develops slowly in the spring but is rarely noticed until more rapid spread occurs at the time of ear emergence or more usually later when environmental conditions tend to be more favourable. Development is most rapid at temperatures of 15–22C; free water is essential for spore germination and high humidity for most stages of infection. Severe attacks at ear emergence have a significant effect on yield. In susceptible varieties losses may be as high as 50 per cent but more usually 5–20 per cent. In most seasons the disease becomes severe too late to be very damaging and in the more resistant varieties it is not important.

Two interesting interactions between brown rust and other leaf diseases have been recorded. A plant affected by mildew is more susceptible to brown rust, and brown rust and *Septoria nodorum* together are much more damaging than the additive effects of the two diseases.

Control

Varieties differ in their susceptibility and this is a factor to be considered in selecting varieties in the southern half of England and Wales where the disease is important. The form of resistance known as 'slow rusting' is now being used by breeders as it is thought that varieties with this character should maintain satisfactory control of the disease. It is also thought that this form of resistance will prove to be more durable than the very high levels of resistance associated with specific genes.

Some fungicides will control brown rust. Because the disease is only occasionally important routine prophylactic treatment is not recommended except where a very susceptible variety is grown in geographic areas where environmental conditions are known to favour the disease. In this case two sprays applied at flag leaf emergence and 2–3 weeks later, just after ear emergence, usually give good control even of a severe epidemic. Criteria for spray applications under other conditions have not been devised for Britain but it is suggested that if the disease is seen on a susceptible variety before the watery-ripe stage a fungicide should be applied immediately; if the first spray is applied before ear emergence and conditions favour spread, a second spray two weeks later may be necessary. In the mid-western states of the USA, where the disease can be serious, a system has been devised for predicting grain loss in a particular crop so that fungicides are applied only if it is economic to do so.

As in the case of yellow rust it is likely that a good control of brown rust is obtained incidentally from the widespread use of broad spectrum fungicides applied in the period from flag leaf emergence to flowering but aimed principally at other diseases.

Black stem rust (*Puccinia graminis*) (Plate 33)
Black stem rust is a serious disease of wheat in North America and elsewhere but in Britain it is rarely important. It is recorded in isolated crops of wheat in most years in south-west England

Plate 33. Black stem rust of wheat, showing the pustules which burst through the epidermis and occur mainly on the stems and sheaths. National Institute of Agricultural Botany.

and in Wales but it is usually too late to be damaging. In the last forty years it has caused crop losses in the south-west on five occasions, the last in 1955. It also affects oats, is rare on barley and rye and fairly common on several grasses. The fungus lives part of its life history on its alternate host the common barberry, *Berberis vulgaris*.

Black stem rust is first noticed on wheat as red-brown pustules containing uredospores which occur predominantly on the stems and leaf sheaths though they also occur elsewhere on the plant. Apart from the occurrence of the pustules on the stem, this rust may be distinguished from other rusts by the torn epidermis which forms a frill around each pustule. The disease spreads in wheat by means of wind-borne uredospores but later in the season black pustules are produced containing teliospores and these give the stem its black appearance. The teliospores remain dormant on stubble and debris throughout the winter and in the spring germinate to produce basidiospores which can only infect barberry. On the barberry the fungus produces the cluster-cup (aecidial) stage and aecidiospores formed can only infect wheat, thus completing the life cycle.

In England and Wales the fungus is apparently not carried over on wild barberry though this does occur in Ireland and is responsible there for local outbreaks of black stem rust in wheat. In England and Wales outbreaks in wheat occur at scattered places in southern and particularly south-western England, initiated by wind-blown uredospores from north Africa or continental Europe, mainly Spain and Portugal, but sometimes France. For an epidemic to occur in June or July when the crop is not yet mature, a plentiful supply of uredospores of strains which can attack wheat varieties growing in southern England is necessary, together with temperature and humidity conditions favouring infection and disease development. Such conditions seldom coincide and consequently epidemics are rare.

P. graminis exists as a number of specialised forms. Each of the forms is further divided into a large number of strains (over two hundred for the wheat form), differentiated on the basis of the varieties which they attack.

Because epidemics are so rare and largely confined to south-west England no control measures are recommended in the United Kingdom. When epidemics do occur they are usually late and for this reason the late-maturing spring wheats tend to be most severely affected. In the 1955 epidemic yield losses of over 50 per cent were recorded.

There was an interesting and unusual sequel to the 1955 epidemic. A small area of wheat is specially grown in the south-west to produce straw for thatching. Thatches prepared from straw harvested in 1955 showed signs of collapse only five years later and this was associated with the use of straw affected by black stem rust. Apparently the rust pustules which burst through the stem destroyed its water-proofing properties. The straw took up water which encouraged secondary rotting organisms, resulting in collapse of the thatch in patches.

Mildew (*Erysiphe graminis*) (Plates 34–6)
The fungus which causes powdery mildew of cereals and grasses is an obligate parasite, that is, it grows only on living green plant tissue. Specialised forms exist, each being restricted to its own particular host so that the forms on wheat only affect wheat and not other cereals, and forms on these other cereals do not affect

Plate 34. Powdery mildew of wheat. This scanning electron microscope photograph of part of a pustule shows the mildew fungus growing over the corrugated leaf surface and producing chains of conidia (spores). Rothamsted Experimental Station.

Plate 35. Powdery mildew of wheat. The white fluffy pustules are most prominent on the leaf blade.
National Institute of Agricultural Botany.

Plate 36. Powdery mildew of wheat. The black spore-cases (cleistothecia) embedded in the white pustules produce ascospores and are an important means of the fungus surviving the harvest period.
Crown copyright.

wheat. (Experimental work has shown that this is not strictly correct and that under certain conditions the barley form may attack wheat. The significance of this possibility in the field is not known but for practical purposes and especially in relation to epidemics of mildew, it can be assumed that the fungus is specific to its own particular host.)

Symptoms

The first symptoms of mildew are small white fluffy pustules growing on the leaf surface. These pustules may remain separate or they may run together to cover substantial areas of the leaf surface. They bear spores (conidia) which are windborne and are the means by which the disease spread. With age the pustules become brown and during the summer bear small black spore-cases (cleisothecia) partly embedded in the fungal felt, which contain a second kind of spore (ascospore). The disease is not likely to be mistaken for any other except possibly when conditions do not favour normal development of the pustule due to genetic, physiological or environmental factors. The fungus may then produce only small dark brown-black marks in the leaf. Some of these usually bear a few threads of the white fungal mycelium but a hand lens or microscope is needed for certain diagnosis.

Wheat mildew affects the ear as well as the leaf blade and sheath. Mildew on the ear goes through the same stages as on the leaf except that it becomes brown more quickly and the black cleistothecia are often seen soon after infection. In the early stages, while the crop is still green, severe ear infection may show in the crop as large discoloured patches. Ear mildew can be distinguished from some other diseases by the ease with which the fungus mycelium can be rubbed off the glume (chaff) although the glume may remain discoloured.

Spread of disease

Mildew survives the winter as white-brown pustules on the lower leaves and, more frequently, on the leaf sheaths. Except in very severe conditions some spread takes place during the winter, albeit slowly, but with the warmer weather in spring spores are produced in large numbers and the disease spreads faster—at first to the neighbouring plants and then, as spore numbers increase, to adjacent wheat crops. The disease is usually most severe on the lower older leaves but new leaves are infected as they are produced and these and the ears can be severely affected.

The disease is not usually evident in the autumn, except on very early-sown crops, and only a small proportion of crops have significant levels of mildew in the spring. However, in the later growth stages, and especially at about the time of ear emergence, the disease may build up.

The cleistothecia are formed from June onwards, usually after a spell of warm weather. Pustules on which cleistothecia are formed stop producing spores (conidia). Cleistothecia serve two purposes. First, they represent the sexual stage in the life history of the fungus and are therefore the means by which recombination of genetic material can occur, to produce new races of the fungus. Secondly, the cleistothecia are the main means by which the fungus survives the period before and during harvest when there is little or no green plant material available.

Ascospores are released from cleistothecia on the crop debris and infect plants in early-sown winter crops and more commonly, volunteer plants which are often present in large numbers a few weeks after harvest. Infected volunteers are the main source of the disease for the later-sown crops. Ascospores are released following rain, dew or periods of high humidity—conditions which also favour the germination of shed grain and the growth of volunteers—mainly in July to September but in small numbers up to November. They are not, however, a source of infection in the spring. The fungus can of course survive the harvest period on any late tillers which survive.

Factors affecting epidemics

For the development of mildew epidemics three inter-related factors must be considered: the fungus, the state of the host (crop) and the environment.

Infection in the autumn, as mentioned above, is usually by conidia from volunteer plants, with ascospores from cleistothecia on straw as an additional source. The fungus is present in most wheat crops throughout the winter and spread can take place by conidia from within the crop when favourable conditions occur in the spring. The higher the level of infection in the spring the greater is the potential for rapid spread. The source of mildew for spring wheat is the winter wheat crop.

There is considerable variability in the plant reaction to mildew, depending on variety and the physiological state of the plant. Wheat is most susceptible during periods of rapid growth, usually in April and May for winter wheat and about a month later for spring wheat. The very significant effect of host physiology on disease development can sometimes be observed on the same plant at the ear-emergence growth stage. The more mature primary tillers may be relatively free from mildew while later-formed and physiologically less mature secondary tillers may be severely affected.

Nitrogen applications increase the susceptibility of the host and late applications prolong the period of susceptibility. Nitrogen applied early in the season, which increases tiller number, tends to increase mildew in the crop. Phosphate and potash, provided they are in adequate supply for normal crop nutrition, do not affect susceptibility to mildew. Dense vegetative growth often favours the disease but this is not related to high seed rates as is sometimes supposed, but to the high nitrogen applications which are often associated with the high seed rates when seed and fertilisers are combine drilled.

Of environmental factors, the most important is temperature. Infection and development of pustules can take place over a wide range of temperatures but the best development of mildew occurs in warm weather (15–22°C); it is retarded at high temperatures (25°C and above). High humidity, though not necessary for spore germination, is important for some stages of the infection process but probably only for short periods. Water occurring as a film over the leaf surface inhibits spore germination. Pustules wetted by rain need several days of favourable weather before spore production recovers to the pre-wetting level.

Severe attacks of mildew may be expected when optimum conditions for all three factors occur at the same time, i.e. the host is growing rapidly or is responding to nitrogen fertiliser, the weather is warm and there is a plentiful supply of spores. In winter wheat, which grows over a long period of time, this means that a single crop may be subjected to more than one severe attack. Attacks may occur in the autumn, though this is unusual in wheat; in the spring; soon after flag leaf emergence; or on the ears. Usually the more severe and damaging mildew attacks develop from the time of flag leaf emergence.

Effect of mildew on the plant and on yield

Mildew affects the plant in several ways. It diverts food materials from the plant to the pathogen and reduces the amount of photosynthetic (food-producing) tissue by causing premature leaf death. It adversely affects the physiology of the plant, for example by increasing the rates of respiration and transpiration. These changes in turn affect the development of the plant in several ways: severe early attacks reduce the number of fertile tillers and the size of the ear; they also reduce the size of the leaves which in turn may affect the size of the grain. Later attacks, after the size of the plant has been determined (the most common in Britain), largely affect the size of the grain, though severe attacks may also reduce the number of grains because some do not mature or are so small that they are lost during harvest. A further important effect of mildew, especially associated with attacks in the early growth stages, is the reduction in the size of the root system which in turn can have serious effects on yield especially if crops are subjected to soil moisture stress. This effect is particularly significant when spring-sown crops are affected as seedlings.

Grain yield losses are related to the severity and earliness of mildew infection and are conditioned by other factors, especially soil moisture. Estimates of annual yield losses in England and Wales based on ADAS survey data, 1970–80, averaged 3 per cent (range 1·5–4·4 per cent) and the disease was the most consistent cause of yield loss of all the leaf diseases. Losses in severely affected crops have been as high as 40 per cent. Yield losses have been related to the severity of mildew at the ear-emergence growth stage in the formula $L = 2\sqrt{M}$ where L is per cent loss and M is the percentage leaf area affected by mildew.

Control

The best means of control is through the use of resistant varieties. Varieties with specific (major gene) resistance which show immunity or near-immunity to mildew give the best control, but such varieties are likely to break down soon after being widely grown and varieties showing a high level of general resistance are more reliable (see NIAB Recommended Varieties). Diversification schemes employing complementary genetic resistance have been devised to reduce the risk of field-to-field spread; appropriate variety mixtures also will reduce spread (page 22). Wheat mildew is not severe frequently enough in currently available varieties for their mildew resistance rating to be a dominant factor in selecting varieties. Furthermore environmental factors affecting the susceptibility of the host are often relatively more important than varietal resistance in determining the severity of the disease in a particular crop. Nevertheless farmers would be wise to avoid growing large areas of known susceptible varieties.

Control by attempting to eliminate sources of infection will not prove entirely satisfactory, partly because spores are carried considerable distances by wind. Nevertheless, it would be unwise to expose crops unnecessarily to large sources of infection. Thus in the case of winter wheat, volunteer wheat plants should be destroyed before the winter crop emerges; in the case of spring wheat, sowing in land adjacent to winter wheat should be avoided. In this way the risk of severe early infection is at least reduced. Although nitrogen increases the susceptibility of plants to mildew it is unwise to modify nitrogen fertiliser programmes below the optimum because a reduction in nitrogen may have a more serious effect on yield than infection with mildew. Early application of nitrogen, before the end of tillering should be avoided unless it is essential for satisfactory crop growth. Such applications encourage the production of the late and highly susceptible secondary tillers.

Some fungicides give an effective control of wheat mildew. Mildew which occurs in the spring and before stem extension in winter-sown varieties is usually not worth controlling unless it is common on the leaf blades. However, mildew at this stage in spring-sown varieties should be controlled as soon as pustules are noticed on the majority of plants. From the period of flag leaf emergence a fungicide should be applied as soon as mildew is seen on the leaves, usually the older ones. One application of an effective fungicide at this time usually confers adequate control for the rest of the season but occasionally a second application may be necessary if further spread is noticed. Sprays applied after the beginning of flowering for the control of mildew are not economically worthwhile.

Strains of mildew resistant to the formerly effective 'DMI' fungicides (see page 26) had become common by 1985. These were associated with only a partial failure of disease control but the level of control obtained in severe outbreaks was not acceptable. Where there is a risk of severe mildew effective fungicides with a different mode of action are recommended.

Septoria diseases

Leaf spot (*Septoria tritici*—perfect stage *Mycosphaerella graminicola*) and **glume blotch** (*Septoria nodorum*—perfect stage *Leptosphaeria nodorum*) (Plates 37–41, page 126–7)

Leaf spot (sometimes speckled leaf blotch) is the common name of a disease caused by *Septoria tritici*. A leaf disease is also the most common symptom associated with *Septoria nodorum*, the cause of glume blotch. In Scotland *S. avenae* f. sp. *triticea* is commonly found causing a disease similar to that caused by *S. nodorum*. The importance of diseases caused by these fungi has only been recognised in Britain since the 1960s. In some districts in the western half of the country they have been a main limiting factor to the growing of winter wheat.

Although grouped together the diseases caused by the two main pathogens are distinct and cause symptoms which are usually readily distinguished.

Symptoms caused by Septoria nodorum

S. nodorum is seed-borne and the fungal mycelium grows from under the seed coat to infect the sheath (coleoptile) very soon after it emerges. Subsequently leaf sheaths and leaves may become infected. Infected seedlings may be stunted and under some conditions they are killed. *S. nodorum* is one of the more important causes of seedling blight in wheat.

During the winter months and more commonly in the early spring *S. nodorum* causes a brown, usually oval, leaf spot though this is rarely conspicuous. During and just after the period of stem extension, the leaf symptoms are uncommon and the fungus is usually found only on the older, often senescent, leaves. After ear emergence the disease may appear on the upper leaves causing brown lesions, generally oval in shape. Later as the disease spreads, the spots may be larger and run together to form irregular areas sometimes occupying a large part of the leaf. Also at this time the disease may appear as small irregular brown spots peppered at random on the leaves as a result of large numbers of individual infections. The disease usually does not become prominent until fairly late in the season, after mid-June and often at the milky-ripe stage or later. However, in severe epidemics, symptoms can be found on the upper leaves soon after ear emergence.

Another characteristic symptom, often noticed on isolated plants, is a papery brown discolouration of the leaf sheath, extending 2 cm or more down the sheath from its junction with the leaf blade.

S. nodorum is also associated with a shrivelling of the nodes though this symptom is normally noticed only towards the end of the season.

The leaf disease is often difficult to diagnose, especially when it is not severe. This is because the symptoms are confused with leaf spots, leaf tipping and other irregular forms of leaf necrosis, often associated with non-parasitic factors, that are fairly common on wheat leaves. Sometimes these dead areas are colonised by fungi including the sooty moulds (page 145) and occasionally by

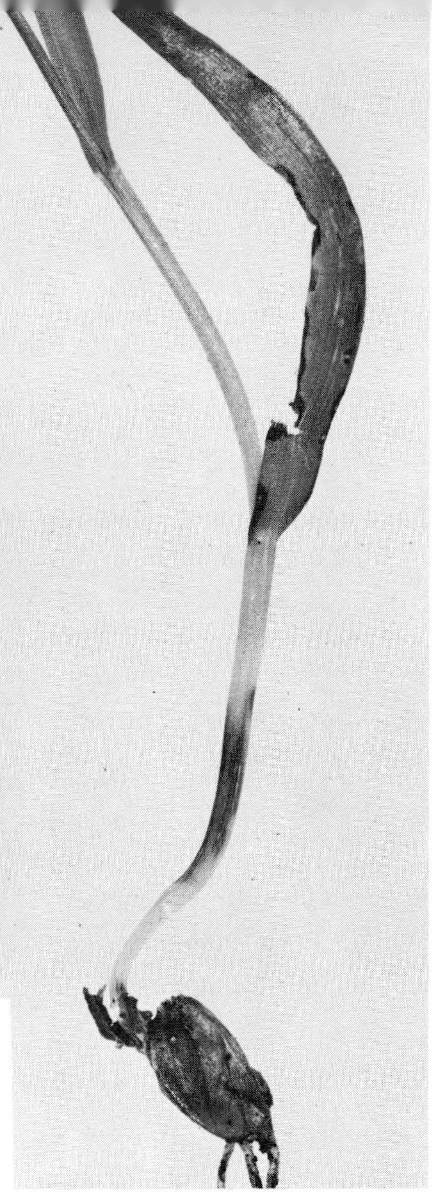

Plate 37. Septoria nodorum: seed-borne infection causing a lesion on the stem and first leaf of a wheat seedling.
Department of Cryptogamic Botany.
University of Manchester.

Plate 38. Septoria nodorum causing well defined spots and irregular disease areas on the leaf blades of a wheat seedling.
National Institute of Agricultural Botany.

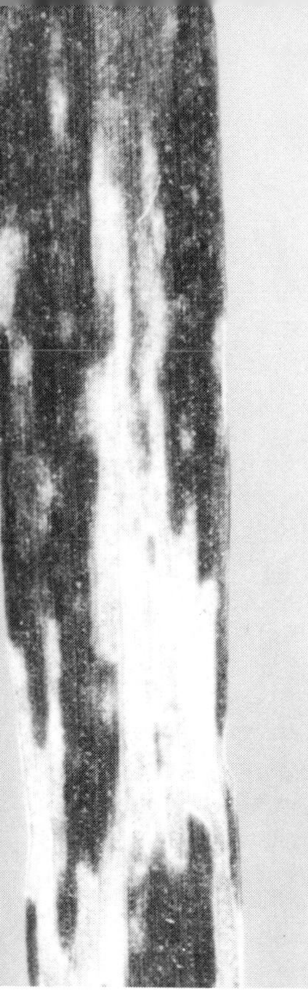

Plate 39. Septoria tritici: Severely diseased wheat leaf bearing spore cases (pycnidia).

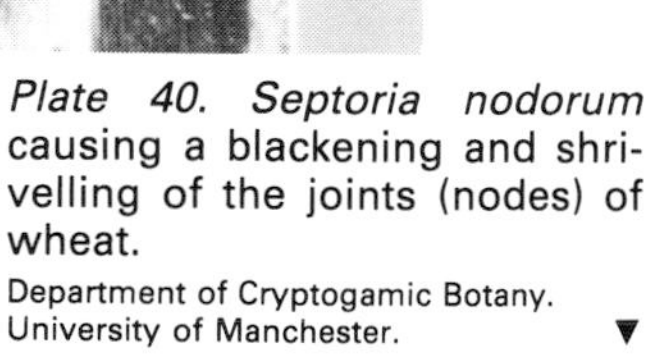

Plate 40. Septoria nodorum causing a blackening and shrivelling of the joints (nodes) of wheat.

Department of Cryptogamic Botany. University of Manchester. ▼

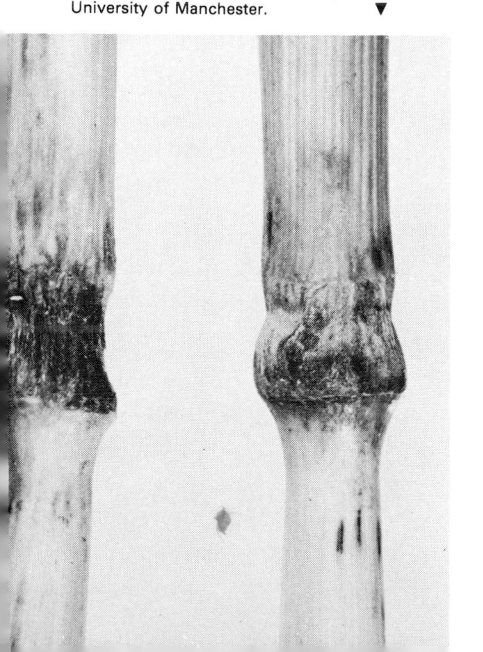

Plate 41. Barley leaf spot caused by *Septoria nodorum* showing the black spore cases (pycnidia) in the centre of the lower spot.

Department of Cryptogamic Botany. University of Manchester.

Asochyta graminicola which is characterised by the production of black pycnidia (spore cases). However, this latter fungus is only a weak parasite and it is usually not damaging to the crop. After the milky ripe stage wheat leaves may be colonised by numerous weak pathogens or pathogens that infect only under special circumstances and these may also be confused with damage caused by *Septoria* spp (see pages 125 and 132).

As an aid to diagnosis the lesions should be examined for the presence of the spore cases (pycnidia). These are often difficult to find especially during the period of stem extension and in the early stages of leaf infection. They are best seen by holding the leaf against the light and examining the lesions with a hand lens. The pycnidia appear as small round pale brown or pink translucent bodies. This is in contrast to the black pycnidia of *S. tritici*.

S. nodorum also causes glume blotch. The most obvious symptoms can be seen on the green ears when dark purple-brown spots develop on the glumes, often beginning at the tips. At this stage glume blotch is seen in the crops as purple-brown patches of variable size often in areas of particularly good growth. As the ears ripen the spots are less obvious; they sometimes develop a whitish sheen and the small black spore cases (pycnidia) may develop, embedded in the glume tissues. Glume blotch can be distinguished from mildew (page 118) and from the secondary black moulds (page 145) which may develop later, by the absence of fungal growth on the surface of the glume. In slight attacks only a few glumes are affected but in severe attacks all parts of the ear including the rachis (stem) may be affected.

Symptoms caused by Septoria tritici

S. tritici is not seed-borne, or only rarely so, and does not cause a seedling blight.

During the winter and especially in the early spring the fungus causes a leaf disease which shows as pale brown spots or frequently as quite extensive spreading lesions which cause a general leaf yellowing in the crop. The lesions bear the conspicuous black spore cases (pycnidia). Later during stem elongation and particularly after flag leaf emergence the disease appears as elongate rectangular lesions or short stripes which are grey green at first but later brown, often a reddish brown. The lesions bear the characteristic black pycnidia and are surrounded by a yellow chlorotic halo. When attacks are severe the lesions merge to form large irregular necrotic areas. Leaf symptoms caused by *S. tritici* are easier to diagnose than those caused by *S. nodorum* and they are not likely to be confused with symptoms due to other causes.

S. tritici has been reported occasionally as a disease of the glume and may be recognised by the arrangement of the black spore cases in lines parallel to the veins.

Sources and spread

Infected seed and straw debris lying on the soil surface from a previously infected wheat crop are probably the main sources of *S. nodorum*, though barley (page 198) and grasses (senescent leaves of which are most susceptible) may also be sources. Sources of *S. tritici* have not been so fully investigated but straw debris and grass hosts are thought to be most important. The perfect stages of both fungi have been found on stubble of infected crops but it is not known whether the ascospores produced by this stage play an important part in the spread of the disease in Britain. In New Zealand, the perfect (*Mycosphaerella*), stage of *S. tritici* is considered important in the epidemiology of the disease caused by this pathogen and in the United Kingdom it may be associated with the very widespread infections which occur in the autumn.

Spores of *Septoria* are produced under conditions of leaf wetness and high humidity. Spread is thought to occur mainly during heavy rain when large raindrops splash the fungal spores from the lesions on the lower leaves to the upper leaves. Some spread may also occur from the combined effects of leaf wetness and wind. The spread of spores can be affected by the structure of the crop. Many modern crops have dense plant growth and this may have a restricting effect on the movement of spores to the top leaves. Although most spread occurs in water splashes to nearby plants it has been shown that spores can become airborne in small droplets, so that the pathogen can be dispersed from field to field.

Infection requires leaf wetness or very high humidity with more prolonged periods for *S. tritici* (c. 20 + hrs) than for *S. nodorum* (c. 3 + hrs). Infection can take place over a wide range of temperatures; the optimum is 15–25°C with *S. tritici* more active in the lower range and *S. nodorum* in the higher range. *S. tritici*, but not *S. nodorum*, can cause infection at temperatures lower than 7°C. These observations help to explain the occurrence of the two species in crops; during the winter and early spring long periods of leaf wetness frequently occur and this, together with the fact that *S. tritici* is not inhibited by lower temperatures, accounts for the fact that this species is then found more commonly; later in the season temperatures rise and periods of leaf wetness and high humidity are generally shorter, tending to favour infection by *S. nodorum*. Infections caused by *S. nodorum* may be arrested by periods of warm dry weather but if wet conditions return the fungus may become active again and cause lesions. Under average conditions, the period between infection and the lesions appearing is 10–30 days, with slighly longer periods for *S. tritici*. Thus infections which occur on the flag leaf soon after it emerges may not become obvious until some time after ear emergence.

High levels of nitrogen fertiliser appear to favour disease caused by *S. tritici* but not that caused by *S. nodorum* which, according to recent observations, may even be depressed by such levels.

Effect on the crop

The diseases are common in the wetter western half of England and Wales, and in some districts are a main limiting factor to the successful growing of winter wheat. In the more traditional arable areas of the east, *Septoria* leaf diseases are less common and are not often serious, but they may be important in wet summers and in some coastal districts.

Both *S. tritici* and *S. nodorum* as leaf and stem pathogens can cause yield losses of up to 25 per cent and occasionally more. When *S. nodorum* causes severe glume blotch in addition, considerably higher losses, up to 50 per cent, may occur. Infection during the period up to flag leaf emergence is rarely severe enough to affect yield. The critical period for effect on yield is from flag leaf emergence until the beginning of flowering when the upper leaves and sometimes the ears are affected. Infection after this has little effect on yield. Because *Septoria* diseases become severe relatively late in the season their main effect is on grain number and especially on grain size. Severe attacks result in much shrivelled grain. Experiments have shown a good relationship between the severity of disease at the milky-ripe stage and loss in yield: each 1 per cent increment in the area of the flag leaf affected by *Septoria* was accompanied by a 1 per cent loss of yield.

In the ADAS surveys (1970–80) *Septoria* diseases caused an estimated annual average yield loss in England and Wales of 2 per cent (range 0·7–7·4 per cent). The disease caused mainly by *S. nodorum* was common and often severe in the wet, season of 1972, and to a lesser extent in 1971 and 1973, but was of negligible importance in 1974–7. In 1979–81 levels increased and in 1981, when May and early June were unusually cool, dull and wet, some severe epidemics occurred especially associated with *S. tritici*. Subsequently severe disease has been mainly associated with *S. tritici*, particularly in the wet summer of 1985 when national losses due to this disease were estimated to be about 8 per cent.

There is some evidence of an interaction between *Septoria* leaf diseases and other diseases. In south-west England severe infection of the upper leaves by *S. tritici* has been noticed on plants affected by yellow rust, take-all or barley yellow dwarf virus and in the Netherlands it has been shown experimentally that *S. nodorum* and brown rust together are much more damaging than the additive effects of the two diseases.

Control

Varieties differ widely in their reaction to both species of *Septoria*. Although none is sufficiently resistant to give a satisfactory control when conditions are favourable for spread of the diseases, some are so susceptible (especially at present to *S. tritici*) that they should be avoided in areas prone to the disease. Apart from differences in inherent resistance, shorter-strawed varieties tend to suffer more severely than the taller varieties because the important upper leaves are closer to sources of the disease at the base of the plant and become infected earlier. Varieties react differently to the two species of *Septoria*.

Seed disinfection with organomercury or other effective fungicides will prevent serious damage to seedlings by *S. nodorum* although it may not prevent some seedlings becoming infected from seed. Suitable cultivations may bury infected straw but stringent precautions aimed at removing this and other known sources have so far failed to prevent some infection occurring on young plants suggesting that other sources e.g. air borne spores from sources outside the field are important in initiating the diseases in crops.

Fungicides give good control of *Septoria* diseases provided applications are well-timed to protect the upper leaves. They are usually most effective if applied soon after flag leaf emergence. At present it is not possible to predict severe outbreaks of the diseases and satisfactory criteria for applying treatments are not available. The present recommendations, which appear to have been reasonably satisfactory, are to apply a spray after flag leaf emergence either as soon as the disease is noticed on the older green leaves or immediately after a period of weather favourable to the disease. A favourable period has been defined tentatively as 'fourteen days during which there are at least four days each with at least 1 mm rainfall or a single day with 5mm rain. *Septoria* may be difficult to diagnose in the field at this time and if a favourable period occurs it is unwise to delay the application of a fungicide even though the disease has not been recognised. Once the disease is established on the upper leaves significant yield losses are likely and good control is unlikely. If a prolonged spell of favourable weather occurs a second spray may be necessary two or three weeks later but sprays applied after the flowering stage are unlikely to be cost effective.

There is some evidence that following mild wet winters, which encourage the development of *Septoria* diseases, a significant effect on the development of disease later in the season, especially in the case of disease caused by *S. tritici*, can be obtained by the application of an effective fungicide at the same time as the eyespot sprays are applied, at the first node growth stage.

Not all fungicides are equally effective against both species of *Septoria*, and in England and Wales resistance to the MBC fungicides (benomyl, carbendazim and thiophanate-methyl) is common in *S. tritici*.

The sprays aimed at *Septoria* control should contain fungicides effective against both diseases and, preferably, also against other serious pathogens such as mildew and the rusts. Such fungicides are also generally effective against other diseases which may develop in the post-flowering period (see 'late leaf disease' below). Experience in the past ten years has shown that yield responses to such fungicides make them cost effective even when severe disease does not occur.

'Late leaf disease' (Various fungi)

In some seasons, in the few weeks before natural senescence occurs, leaf lesions may develop, especially on the flag leaf and the leaf below the flag. The symptoms range from a peppering of indefinite small brown spots, through larger spots to extensive diseased areas, often restricted to certain parts of the leaf, for example, the upper surface of the 'bend' of a leaf or the part near the stem. The lesions, which are mainly dark brown but occasionally pale brown to white, develop after flowering and are often associated with a heavy deposit of pollen. This association appears to be significant, but the same areas of the leaf are also those where moisture tends to collect and persist and this factor may also be concerned with the development of the diseased areas.

Various fungi have been isolated from the diseased tissues but the most common from well-defined lesions are *Botrytis cinerea* and *Alternaria alternata*. Both are regarded as weak parasites of otherwise healthy tissues but, as has been shown in other crops, infection may be stimulated by the pollen present on the leaves to cause severe lesions.

Several other fungi cause lesions that develop late. Lesions caused by *Septoria* spp. can usually be distinguished by the presence of pycnidia (spore cases, see page 124). *Didymella exitialis*, which is very common on cereal straw and is usually considered to be a colonsier of dead or dying leaves in wet seasons, can cause brown lesions on otherwise green leaves; its black spore cases (in this case perithecia) may be confused with the pycnidia of *Septoria*. *Fusarium* spp., especially *F. nivale* (page 95), can also cause leaf spots late in the season. *Ascochyta graminicola* produces black pycnidia on dead tissue, although this fungus is more often seen earlier in the season. Dead tissues late in the season may be colonised by the sooty moulds (page 145). Although a single pathogen may sometimes be responsible for the damage described, very often, and especially in wet summers, several fungi may be identified in different parts of the diseased areas.

These late-season diseases are less damaging to yield than diseases that occur earlier but there is some evidence that they may reduce yields especially in seasons when there is an extended period for grain filling. Although no specific fungicide treatment can be recommended for these diseases, which occur erratically, they are well controlled by some of the broad-spectrum fungicide mixtures applied for the control of late diseases, especially *Septoria* (page 131).

Cephalosporium stripe (*Cephalosporium gramineum*)

This disease affects cereals and grasses but in Britain it is mainly a disease of wheat. Affected plants occur at random in the crop though there is sometimes a tendency for the disease to be more common on the headlands.

Diseased plants are usually noticed in June or later when the upper leaves have wide pale green or yellow stripes extending the length of the leaf blade and down the sheath. There is frequently only one, sometimes two or three, stripes on each of the leaves and affected plants are very conspicuous among healthy green ones. If the stem is cut across, the vascular strands in the stem and sheath are seen to be stained dark brown. The yellow striping symptom does not last long in the field since the plant is usually killed before the stem has achieved its full height. For this reason affected plants are soon hidden by their taller, healthy neighbours and are often overlooked. It is a characteristic of attack by this disease that virtually no grain is produced in the affected ear. This feature, coupled with the development of a watery-brown band up to 1 cm deep, immediately underneath the top stem node or joint (and often under some of the lower nodes), and the discolouration in the nodes when cut across, serve to distinguish this disease from the several other causes of 'whiteheads'. In most cases only small numbers of plants are affected and the disease is rarely of economic importance.

The fungus is soil-borne and invades the plant through damaged roots or other underground tissues and grows mainly in the water-conducting (xylem) tissues. Roots may be damaged naturally as occurs during root emergence or, more significantly, due to feeding by pests. Cephalosporium leaf stripe is most common in wheat after grass when pests may be present in large numbers and the most severe attacks have been associated with early damage by wireworms. Slug damage has also been associated with some severe attacks. The fungus can survive in crop debris for at least one year and also in grass hosts. Seed-borne infection has been recorded but this is regarded as an unimportant source of the disease.

Barley yellow dwarf virus (BYDV)

Wheat is more tolerant of this virus than barley and oats but it can suffer severe attacks. In the epidemics of 1981, wheat was affected as frequently and damaged as severely as barley, and yield losses were equally serious. The most severe attacks occur in autumn-sown wheat particularly in crops which have been sown early (i.e. before mid-October) or where sowing has followed soon after ploughing a ley where the aphid vectors have moved within the field from unrotted grass infected with the virus to the young wheat plants. In such attacks especially when the more severe strains of BYDV are also involved, affected plants may be killed during winter but usually they survive until the spring when they appear stunted with some leaf yellowing and occur in patches which are usually small and well defined but occasionally extensive. More commonly symptoms due to infections in late autumn and winter are seen in late May and early June at flag leaf emergence. Such plants may be slightly stunted but the main symptom is the colour of the leaves. The upper leaves of affected plants, notably the flag leaf, are pale yellow, sometimes with a touch of pink extending from the tip of the leaf towards its base. Again the affected plants occur in distinct patches or occasionally in more extensive areas. When spread continues in early spring the symptoms tend to appear somewhat later and affected plants occur in less well defined areas.

Plants infected early in the autumn with the more severe strains and which are killed out during the winter period rarely occupy significant areas and losses may be compensated by adjacent healthy plants. Usually plants survive to produce an ear but with the more severe strains and early infections, yield can be reduced by 50 per cent or more and the grain is shrivelled. This is in contrast to winter barley where plants affected under similar conditions may remain stunted and produce infertile tillers. Later infections or infections with the less severe strains have corresponding less effect on yield. The incidence of BYDV in winter wheat has tended to be lower than that in winter barley, possibly because of the generally later sowing date. The overall effect on yield was rarely severe and even in 1981, when some of the worst attacks in winter wheat occurred, the average loss in infected fields was probably not more than one per cent. In more recent years winter wheat crops have been drilled earlier and occasionally some very severe attacks of BYDV have occurred. However the earlier drilling dates have also been accompanied by a general use of insecticide in the autumn to control the aphid vector. The full potential of the disease on these early drilled crops has therefore not been observed.

Apart from the symptoms the epidemiology of the disease and its control is the same in wheat as in barley and is more fully described on page 200.

European wheat striate mosaic

This is a very damaging disease which can also affect other cereals and ryegrasses. Infected plants have white-yellow narrow stripes on the leaves. Plants affected early usually die and those affected later produce 'whiteheads' and ears with no grain or shrivelled grain. The causal agent is not known but is possibly a mycoplasma. It is spread by the planthopper *Javasella* (*Delphacodes*) *pellucida* usually from adjacent grass fields and hedgerows. Fortunately agricultural practice and climate in Britain do not favour spread of the disease and economic losses rarely occur.

Manganese deficiency

The first sign of manganese deficiency in wheat crops may be noticed in May when pale-coloured areas of variable size can be seen from a distance. The plants in such areas are pale green and then develop very small white or pale brown spots. Later the spots may coalesce to form lines on the leaves, sometimes concentrated in the middle of the leaf, especially in older ones. Young plants may have a floppy appearance. Later in extreme cases up to half of the leaves, especially the older ones, can be killed and the plant is much reduced in vigour, with consequent serious yield losses. Manganese deficiency is most common on peaty soils or on mineral soils with a high organic content, particularly those with a high pH. Some varieties are more sensitive to the deficiency than others.

Control measures are the same as for the deficiency in oats (page 238).

Bunt (*Tilletia caries*) (Plate 42)
Until organomercury seed disinfectants came into general use, bunt (or 'covered smut') was a serious disease in England and Wales. It is not an easy disease to recognise in the field but near to harvest the experienced eye can distinguish the rather staring ears, whose glumes are pushed further apart than in the healthy ear by the rounder, plumper 'bunt balls' which replace the grain. Usually all the grains in an ear and all the ears of one plant are infected. Each 'bunt ball' or 'butt' is a bag full of the black, fishy-smelling spores of the bunt fungus. The disease was called 'stinking smut' at one time.

During harvest, as the grain is threshed, most of the 'bunt balls' are broken and the spores distributed over the surface of the healthy grain. If the crop is heavily infected, it may be made totally unfit for milling because of blackening of the grain, resulting from a heavy dusting of spores concentrated at the brush end of the grain. The combine, trailers, sacks, etc., are also contaminated so that grain from subsequently harvested clean crops is also at risk.

Contaminated grain has been used as feed for cattle, pigs and poultry without apparent ill-effect.

Contaminated grain, if sown without seed disinfection, gives rise to infected seedlings which are, on the whole, indistinguishable from healthy ones since their growth is virtually unaffected until after ear emergence. Sometimes the affected plants are slightly shorter than healthy ones. As the grain germinates after sowing, so do the spores of the fungus, which penetrate the young shoot at a very early stage, before emergence. From then onwards the mycelium of the fungus keeps pace with, and only just behind, the growing tip of the plant, until ear emergence. As the young grain develops in the ear, it is invaded by the fungus from within the plant and 'bunt balls' are formed in place of grain.

Resistance to bunt is known but cannot be relied upon for control, because the fungus exists in many different races or strains so that it is doubtful if resistant varieties would survive for very long. It has not been necessary to put this to the test because of the outstandingly efficient control given by the organomercury seed disinfectants (see page 23) and more recently by other seed treatments.

Another *Tilletia* species (*T. laevis* syn. *T. foetida*) is a common cause of bunt in North America, and yet another species (*T.*

Plate 42. Bunt of wheat. The diseased grains (above) are filled with black oily spores which are dispersed during threshing. National Institute of Agricultural Botany.

controversa) causes dwarf bunt in both North America and Europe. Neither has been recorded in Britain.

Dwarf bunt causes excessive dwarfing and tillering. It is common in Eastern Europe and has been reported as an increasing problem in Sweden. It differs from bunt in that spores remain viable in the soil for several years so that the disease is not controlled by organomercury seed disinfectants alone, although these will control the seed-borne fungus.

Loose smut (*Ustilago nuda*) (Plate 43)
Unlike bunt (or covered smut as it is sometimes called) loose smut is very easily recognised in the field at heading time because the loose mass of black spores is released as soon as the ear emerges from the boot. These blackened ears are so obvious, scattered through the crop, that the disease appears rather alarming even at the very low levels of infection which are usual in Britain. One infected ear in 1000 is easily seen and a 1 per cent infection, which is not seriously damaging to yield, appears devastating. However, levels of infection as high as 20 per cent are not unknown, so the potential of the disease for serious damage cannot be ignored.

In the field the obvious black ear symptom soon disappears, because the spores of the fungus which cause the disease are quickly blown away or washed off in rain, leaving the bare stem (rachis). Some of the spores that are blown about in the crop alight on the stigmas of open flowers, germinate and the fungus enters the developing grain. Unlike bunt balls, the infected grains appear perfectly healthy, the infection being carried internally and there is no external contamination or discolouration to be seen. There is thus no consequential loss of crop during harvest or from contamination in combine or sack and infected seed is still perfectly acceptable for milling purposes.

If used for seed, however, progress of the disease follows that of bunt (page 136) very closely. The infection present internally in the seed keeps pace with the development of the plant throughout its life, remaining viable just behind the growing point, until the ear develops within the boot. At this stage, the spores of the fungus are formed in place of flower parts and the ear emerges as the familiar sooty spike.

The level of loose smut in a crop reflects the amount of disease in the previous seed crop and the weather conditions during the flowering period of that crop. Cool, wet weather lengthens the time the florets of wheat remain open, and hence the time available for the fungus spores to enter the flower parts and invade the ovary. In view of the influence of weather, it is not surprising that the level of infection varies from season to season. Nevertheless the trend over a period of years will be for the level of infection in a given seed stock to increase.

Since the infection is internal, organomercury seed treatments, which only disinfect the surface of the seed, are not effective against loose smut. The traditional method of eradicating the loose smut fungus was by immersing seed in warm water, presoaked, then at 54°C for ten minutes. This was expensive and was therefore used only in the more valuable seed stocks. Now

Plate 43. Loose smut of wheat. The grains are replaced by masses of black spores. The spores are later blown or washed away leaving a bare stem.
National Institute of Agricultural Botany.

an effective and much more easily managed alternative has been widely adopted in the form of systematically-acting fungicidal chemicals which are applied, usually in combination with other chemicals, for the control of all seed-borne diseases. These chemicals are able to enter the seedlings and kill the smut fungus within the plant without injuring the plant itself. The treatment costs are more than for the standard organomercury treatment and need only be applied to stocks being grown on for seed production.

Farmers normally rely on seed produced under the UK Seed Certification Scheme of the Ministry of Agriculture, Fisheries and Food. EEC regulations impose a standard of not more than one smutted head in 200, which is generally considered insufficiently stringent to keep the disease in check. Most seed produced meets the 'Higher Voluntary Standards' level of one infected head in 1000.

Farmers wishing to save their own seed should remember that smut may increase up to twentyfold in one season, although the actual increase will depend on the weather at flowering, as mentioned above. With such a potential increase in mind, seed crops should be inspected and smutted heads counted in several sample areas at flowering time before a decision is made to use the grain for seed. Alternatively, where there is any doubt about the loose smut status of a seed stock then an effective systemic fungicide seed treatment should be used. The embryo examination used for barley loose smut (page 208) is not suitable for wheat.

Although varieties differ in susceptibility to loose smut, the fungus exists in several forms or 'races' and resistant varieties tend to show higher levels of infection the longer they survive in commerce. This happened in the case of Cappelle Desprez against which a new race of the fungus developed.

Ergot (*Claviceps purpurea*) (Plates 44–5)

Ergots are hard, purplish-black fungal bodies which replace individual grains in the ear. Wheat ergots may be similar in size and shape to wheat grains but most often they are several times larger, protruding obviously from the spikelet. The ergot is formed of a mass of fungal tissue which is watery-white internally in contrast to the purple-black outer covering. The disease is important because the ergots, if eaten, can cause poisoning of humans and animals.

The disease is confined to the ear. Yield losses are negligible but it may be difficult to remove all the ergots from the harvested grain and, with stringent conditions set for the sale of cereals for

Plate 44. Ergot in (left) wheat and (right) barley. The grains in a few spikelets are replaced by dark, hard bodies which protrude from the spikelets.
National Institute of Agricultural Botany.

Plate 45. Ergot. The black ergots germinate to produce pinhead structures (apothecia) which produce spores. These spores infect the flowers of cereals.
National Institute of Agricultural Botany.

flour, feedingstuffs or seed, contaminated samples are difficult to sell or command a much reduced price. It is this aspect of the disease that may cause economic loss to the cereal farmer.

The causal fungus, *Claviceps purpurea* affects cereals, most commonly rye, and grasses. The disease was uncommon in wheat for many years but a significant number of cases occurred during the 1980s. Ergots are less frequent in barley, rare in oats and very common in grasses. There are different strains of the fungus, each with a different host range but ergots on any host should be regarded as a potential source of infection for cereals.

Ergotism

Ergots contain a mixture of alkaloids in varying concentrations, some of which are poisonous to humans and animals. The quantity of ergots that has to be eaten before symptoms of ergotism are produced varies considerably depending on the type and concentration of these alkaloids. There are two main forms of ergot poisoning. One form, mainly associated with the vascular system, causes a constriction of the fine blood vessels resulting in a range of symptoms including nausea, feelings of extremes of hot and cold, a loss of feeling in the extremities and in severe cases a dry gangrene leading to the loss of nails, fingers and toes and sometimes whole limbs. The other form, more closely associated with the nervous system, involves numbness and violent pains, fits and in extreme cases insanity.

Ergotism in humans was common in the Middle Ages when it was known as St Anthony's fire; since the eighteenth century few cases have been recorded in Britain. In this century there appear to have been two cases, both in the 1920s and both associated with eating ryebread. Ergots are much more common in rye and ergotism is still recorded occasionally in countries of continental Europe where ryebread is regularly eaten.

Ergotism in livestock is usually chronic as a result of continued consumption of small quantities of ergot. Most cases are attributable to grazing pastures in which grasses have been allowed to produce mature seed heads, some of which contain ergots.

Some of the alkaloids in ergots are valuable in medicine and are produced specially for this purpose by artificial inoculation of rye.

Spread

At harvest some ergots fall to the ground. Provided they are not buried deeply they germinate the following early summer at about the time cereals and grasses are in flower and produce a variable number of flesh-coloured fruiting bodies resembling drum sticks (Plate 45.) These are up to 2·5 cm in length and in the swollen tips a large number of spores (ascospores) are produced which are carried by air currents or insects to the open flowers of cereals and grasses. The spores are often trapped on the stigmas of the open flowers where they germinate and infect the ovaries. About seven days later a sticky liquid exudes from the infected flower; this is the honey dew stage. The liquid contains numerous spores of a different kind (conidia) which may be carried by insects attracted by the honey dew, or in rain splashes, to other flowers providing an effective means of secondary spread. This phase continues for about two weeks in individual ovaries, occasionally longer, until the secretion of honey dew ceases. Fungal growth then increases to produce the ergot in place of the normal grain.

The germination of wheat ergots and the release of ascospores often does not coincide well with the flowering period of wheat so that ergots arising directly from ascospore infection of wheat may not be common. However, the ascospores can infect grasses and secondary spread from the honey dew stage on these can then occur. Blackgrass (*Alopecurus myosuroides*) has been identified as the most important source of infection of this kind for wheat. Other grasses which may be involved include annual meadow grass (*Poa annua*), meadow foxtail (*Alopecurus pratensis*) and perennial ryegrass (*Lolium perenne*).

Ergots survive in the soil for only one year. They can withstand freezing conditions and in fact require a low-temperature treatment followed by moisture and temperature of at least 12°C to germinate in the spring. The disease is favoured by cool wet weather which not only facilitates the production and dispersal of both types of spore but also prolongs the flowering period during which infection takes place. The number of infections may also be influenced by the flowering habit of the host; rye and the grasses which are most susceptible to the disease have flowers which remain open for much longer than those of other cereals.

Control

It is not normally necessary to apply special control measures. If a serious outbreak occurs and a susceptible crop is to follow then steps should be taken to reduce the number of ergots in the field. This can be attempted by careful ploughing to bury the ergots deeply enough to prevent germination. It is important that the furrow is inverted to ensure that the ergots are covered by at least

20 cm of soil. However, a better method is to employ a rotation. Since ergots survive in soil for only one season, a rotation with a non-cereal crop or grass is an effective control.

Control of grass weeds, especially blackgrass, within crops is most important since this will remove a major source of the disease. Grasses on and around headlands also provide a source of infection and cereal ergots are often numerous on the outer edges of the crop; if possible the grass should be cut before the cereals come into flower. Ergots from grasses can also contaminate a grain sample.

Mechanical cleaning of contaminated samples will remove most of the ergots, including those from grasses, and should enable most samples to attain the standard required for animal feed. Since screenings from such samples will contain a large number of ergots they should be buried deeply or burned and must not be fed to livestock.

The traditional way of separating ergots from grain is by immersion in a brine solution (18 kg common salt in 300 litres of water) when the ergots float and can be removed. The grain is then washed in water and dried.

Chemical control would be justified only in special circumstances, e.g. in hybrid seed production when flowers remain open for long periods and are therefore more susceptible. Some control has been achieved experimentally by frequent fungicide applications during flowering but no recommendation can be made.

Scab (*Gibberella zeae* [*Fusarium graminearum*])

This fungus behaves in a similar way to other *Fusarium* spp. in causing seedling blight and brown foot rot in cereals (page 95) but is distinguished by the symptoms it causes in the ears. Like the other species of *Fusarium* it produces pinkish-red spore-masses on ears (and stems) but among these, blackish spherical bodies are formed which represent the sexual or 'perfect' stage of the fungus (*Gibberella zeae*). The fungus can invade the grain and when this happens, toxic substances (mycotoxins) are produced to which various animals and man are very sensitive. The toxic substances are very stable and persist long after the death of the fungus and can, for example, cause vomiting in pigs fed with infected grains and in humans who eat foodstuffs prepared from infected grain. Fortunately scab is only rarely serious in Britain and then is mainly confined to the western and northern areas of the country.

The same fungus is the cause of foot rot in maize (see page

259), so that introduction of maize, whether for silage or grain, as a 'break crop' entry into wheat carries the risk of increased damage from scab in the wheat crop.

The scab fungus persists in cereal residues in the soil and on seed and no satisfactory control measures can be advised. Some seed treatments reduce, but generally do not eradicate, infection carried by the seed.

Botrytis disease of spikelets (*Botrytis cinerea*)
The disease is seen on very few, rarely more than two or three, spikelets per ear and only on occasional ears. It is particularly conspicuous when the ears are still green. The fungus appears to infect the spikelet by colonising the dying anthers and then growing down into the developing grain or into the adjacent parts of the spikelet, killing the grain and causing large pale brown lesions on the glume. With a hand lens the grey mould of the fungus can usually be seen near the tip of the glume and this may cover the whole spikelet if wet conditions persist. *Fusarium* spp., especially *F. poae* and *F. nivale*, can also cause lesions on the glumes (page 97).

Further information on the disease which also affects barley is on page 209.

Twist (*Dilophospora alopecuri*)
The disease which affects wheat and grasses is of historic interest only. It was once widely distributed in Britain but it is now extremely rare and has not been seen in the field by the writers. The ear becomes infected inside the sheath and this may prevent it from emerging normally, causing the stem and ear to be twisted. The ear is covered with a white fungal mass which on exposure to the air shrivels and blackens and bears masses of the fungal spore cases (pycnidia). The fungus is carried inside the seed as well as on the outside so that seed disinfectants give only partial control.

Black mould (Plates 46 and 47)
Under some conditions, mainly those causing a premature senescence of the plant, ears of wheat (and to a lesser extent oats and barley) become blackened by the growth of black sooty moulds on the surface of the various parts of the ear, especially the glumes. Sometimes only patches of plants are affected but in some seasons very large areas may show the condition. The moulds most commonly involved are *Cladosporium* spp. but

species of *Alternaria* are also fairly common. These fungi are not parasitic or at most are very weak parasites and are certainly not responsible for the thin ears with shrivelled grain on which they are often found. They are common on all kinds of dead plant tissue and their presence on ears should be regarded as a sign that the plants have been affected by a disease or some other adverse factor, perhaps some considerable time before the moulds appeared. They normally occur on ears which have ripened prematurely for any reason. Diseases which may cause premature ripening (whiteheads) include take-all, eyespot, sharp eyespot and

Plate 46. Black sooty moulds (mainly *Cladosporium*) affect the thin wheat ears of prematurely ripened plants. Normal ears on left.
Rothamsted Experiment Station.

Plate 47. Plots in which crop on the right was affected by take-all and the prematurely ripened plants were colonised by black sooty moulds. The crop on the left was healthy.

brown foot rot, and also severe attacks of leaf diseases and barley yellow dwarf virus. Adverse soil conditions, especially drought, herbicide damage and heavy infestation of the ears with aphids, may also be associated with black moulds on the ears, and plants which have lodged are often severely affected.

Black moulds should be distinguished from ear diseases caused by parasitic fungi especially mildew (page 120) and glume blotch (page 128), although attacks of these diseases may be followed by the black moulds acting as secondary colonisers of dead tissue especially towards harvest time in wet seasons.

Grain from crops affected by black moulds usually has a normal appearance but occasionally grain from crops severely affected has a poor appearance which may affect its saleability adversely.

Shrivelling of grain is, of course, associated with some primary cause of damage to the plant and not with black mould. Flour prepared from grain affected by *Alternaria* spp. in particular may be seriously discoloured (see black point, page 152).

As black moulds are a sign and not a cause of trouble, the primary cause of premature ripening should be ascertained. Some causes of 'whiteheads' are on page 83 but adverse soil conditions including poor structure and drought should also be investigated. Appropriate measures can then be considered to avoid a recurrence of the problem.

Some fungicides applied at or after ear emergence for the control of leaf and ear diseases may also give some control of black moulds and occasionally improve the appearance of the grain. However, the control of black moulds on crops severely affected by prematural ripening is likely to be poor. The development of black moulds cannot be predicted and the prophylactic application of fungicides for their control is not recommended. Fungicides applied after black moulds have appeared will not be effective.

Late frost damage

Frosts in May and June may damage leaves and ears though the symptoms may not be noticed until much later and their association with the frosts is often overlooked.

Damage to leaves shows as white or pale brown areas at the leaf tips, demarcated from the rest of the green leaf by a straight line across the leaf. The extent of the tipping varies but up to half of the leaf may be affected and the damage is usually confined to two or three successive leaves. Occasionally frost may damage the lower parts of the stem causing a bleaching or browning of the stem and a dark brown discolouration of the joints (nodes). This damage is revealed when sheaths are stripped away and frequently consists of a very sharp, narrow depression, running on one side between the two lowermost stem joints. The depression may be in the middle of a flattened side of the stem which is often paler in colour than the healthy straw. This kind of injury is often overlooked but has been noticed as a cause of patchy lodging in some years in fen and skirt soils in East Anglia. It can be confused with eyespot. Stems damaged in this way bear poorly filled ears.

Frost can cause 'blindness' or 'deafness' in ears. Affected spikelets are bleached and contain no grain, so they stand out in contrast to the unaffected parts of the ear. The tips of the ears are most commonly damaged but sometimes only the middle or lower parts are affected. Spikelets do not all mature at the same time and only some are sensitive to frost damage at any particular time. Damage often seems to occur during or just after the time of ear emergence.

'Blindness' may also be caused by many other conditions especially soil-borne diseases (e.g. take-all), some pests and adverse soil conditions (e.g. drought, poor drainage). However, in these cases, although most of the blind spikelets are towards the tips and bases of the ears, there is no sharp distinction between affected and unaffected spikelets and other parts of the plant show accompanying symptoms such as leaf yellowing or stunting. 'Blindness' should also be distinguished from Fusarium ear blight (page 97).

Copper deficiency

Copper deficiency occurs on a range of soil types but mainly in East Anglia especially on the lighter peats of the fens and on organic sands, and in the south of England on shallow loams with a high organic content overlying chalk. Wheat, barley and oats can be severely affected but rye is less susceptible.

Early growth is normal but at tillering young leaves become pale green with a yellowing and whitening of the leaf margins and tips. Tips of affected leaves may bend at right angles to the stem and sometimes leaves twist into spirals. Such symptoms are associated with severe damage. Often symptoms are not noticed until after ear emergence when ears may be trapped in the sheath or, if they emerge, the tips are white and contain no grain. An excessive number of late weak tillers is sometimes produced. Mild copper deficiency symptoms may not be noticed until some time after the ears have emerged when patches remain green longer and the spikelets at the tips of ears fail to produce grain. This last symptom should be distinguished from similar symptoms caused by frost and some diseases in a severe form (e.g. take-all, eyespot, fusarium diseases).

In southern England on organic chalk loams a different symptom has been recognised. This appears after ear emergence when a dark discolouration of the stem is seen, first on the upper part of the stem, then under the ear and finally on the lower parts of the stem. This is due to the formation of a black pigment (melanin) in the tissues, and the disorder known as 'blackening' affects large patches of plants which can be seen clearly at a distance. Affected plants are dark and may die. The strength of the straw may be affected so that stems curve and collapse despite the light ears. On the chalk soils copper deficiency symptoms are often most severe in the third or fourth crop after breaking up the original downland turf and in these situations it is most commonly seen in barley.

Copper deficiency can cause severe damage, 80 per cent grain yield losses having been recorded, although average losses are much lower. The grain that is produced is of poor quality.

Control

Copper deficiency can be corrected by spraying the crop at late tillering with 2 kg per hectare of copper oxychloride or cuprous oxide (both containing 50 per cent copper) in 500 litres of water. This spray timing is later than that advised for most herbicides so that the two should be applied separately. The deficiency can also be corrected by applying 30–60 kg per hectare of copper sulphate to the soil before drilling but in a dry season a further response can be expected from sprays applied in mid season.

Loose ear

This is a fairly rare and unusual condition in which bleached ears of isolated plants or, very rarely, small groups of plants are very conspicuous amongst healthy plants. The symptom is more conspicuous than a typical 'whitehead' symptom because only the ear and the stem immediately below it are bleached white and are held by a sheath and flag leaf which are green as is the rest of the plant. The ear and stem can be pulled out easily from the sheath and on detailed examination it will be found that the stem has been severed inside the sheath, usually within 2 cm of the topmost joint.

The cause of this condition is not known although it has been associated with damage during grazing by stray cattle. However, it has been reported under other circumstances and may be due to some abnormality in the relative growth rate of the stem and enclosing leaf sheath.

Herbicide injury (Plates 48a, b, c)

Growth regulator herbicides applied before the recommended time cause a range of symptoms similar to those in barley (see page 211). They include tubular leaves and ear malformations where the relative positions of the spikelets are altered; spikelets may be opposite, more numerous, bunched, etc. Sometimes the stem splits to form branched ears, or the ear is trapped in the flag leaf causing the stem to buckle.

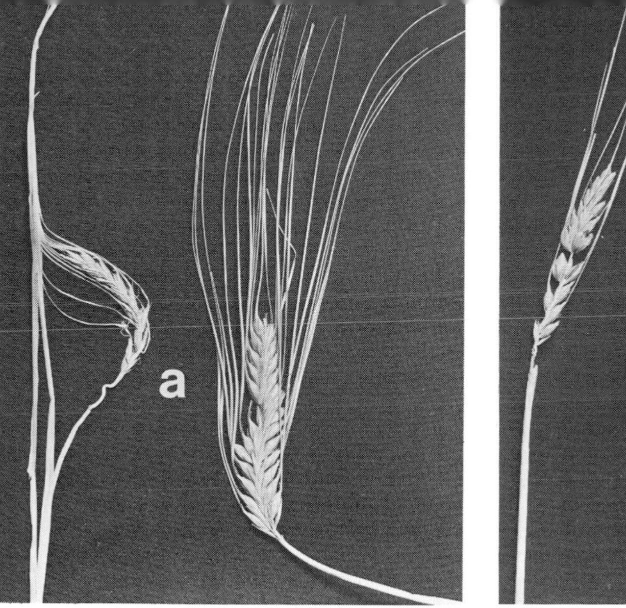

Plate 48a, b. Herbicide damage (a) and (b). Application of a growth regulator herbicide before the recommended time can cause abnormal ears—various forms of distortion in barley. The Boots Company Ltd.

Plate 48c Herbicide damage (c). The effect of late application of a growth regulator herbicide at the jointing stage in wheat, resulting in thin ears subject to black moulds. Wheat sprayed at the correct stage is on the left. The Boots Company Ltd.

Late applications of these herbicides do not cause ear malformation but may result in poorly-filled ears; yield losses of up to 100 per cent can occur in badly affected areas. This damage is more common in winter wheat than in other cereals. At harvest the affected areas present a blackened appearance due to the growth of secondary black moulds (page 145) on the ears. The symptoms are easily confused with those caused by take-all (page 87). In crops where take-all is absent the symptoms can be attributed to the herbicide because the roots remain white and healthy. Where take-all is present diagnosis is more difficult, especially since the effects of the disease are made worse by the herbicide damage. Under these conditions the distribution of affected plants can give a clue as to whether disease or herbicide or both are the cause of the damage. Take-all tends to occur in irregularly shaped patches whereas damage by herbicide follows a pattern determined by the passage of the sprayer; any areas missed by the sprayer appear normal and stand out in sharp contrast.

Black point

Black point describes the sometimes shrivelled, brown-black discolouration of the embryo (germ) end of the grain. In wheat the symptom is sometimes confused with that of 'bin burning' caused by overheating during storage and which, unlike black point, may also affect germination adversely.

Black point can cause a discolouration of the flour (due to dark specks) and for this reason affected wheat samples are not acceptable to millers. Because of the appearance of the grain, affected samples are also not acceptable for seed (though germination is usually normal).

Several fungi including *Cladosporium* spp. and *Botrytis cinerea* have been associated with this symptom but species of *Alternaria* are the most common. They are regarded as saprophytes or weak parasites and are favoured by wet conditions (see black moulds, page 145).

Chapter 4

BARLEY PESTS

PLANT DAMAGE SYMPTOMS

Symptom	Pest
a. *Seed or seedling damaged at or before emergence*	
Seeds missing from drill row	Birds (p. 75) Rats, mice (p. 77)
or hollowed out	Slugs (p. 166) Wireworms (p. 55) Mice (p. 77) Bibionid fly larvae (p. 48)
b. *Shoot develops but does not reach soil surface*	
Shoot short and swollen, root tips club-shaped	Seed treatment injury (p. 23)
Shoot brown and shows signs of feeding	Slugs (p. 166) Leatherjackets (p. 162) Wireworms (p. 55) Wheat bulb fly larvae (p. 157) Frit fly larvae (p. 160)
c. *Seedling or tillering plant damaged*	
i. Whole plants affected, usually yellow first	
Plants pulled out and left on soil	Birds (p. 75) Rats, mice (p. 77)
Plants bitten near soil level	Slugs (p. 166) Bibionid fly larvae (p. 48) Leatherjackets (p. 162) Wireworms (p. 55) Grass moth caterpillars (p. 49)

	Swift moth caterpillars (p. 49) Cutworms (p. 49) Chafer grubs (p. 57) Wheat shoot beetle larvae (p. 66) Millepedes (p. 66)
Plants not bitten	Cereal cyst nematode (p. 167) Frost damage (p. 210) Deep sowing (p. 84)
Shoot swollen or mis-shapen	Gout fly larvae (p. 157)
Plants slightly stunted, leaves spotted or flecked	Aphids (p. 165) Thrips (p. 165)
ii. Centre shoot yellows and dies (deadheart), outer leaves stay green	Wheat bulb fly larvae (p. 157) Frit fly larvae (p. 160) Late-wheat shoot fly larvae (p. 157) Grass moth caterpillars (p. 49) Common rustic moth caterpillars (p. 54) Grass and cereal fly larvae (p. 157) Yellow cereal fly larvae (p. 157) Bean seed fly larvae (p. 42) Wireworms (p. 55)
iii. As above, but a small neat hole often present near shoot base	Flea beetles (p. 164) Wheat shoot beetle larvae (p. 58)
iv. Leaves bitten, often well above ground	
Long narrow holes	Slugs (p. 166)
Roughly circular holes	Grass moth caterpillars (p. 49)

Symptom	Pest
v. Leaves neatly cut off	
Horizontal bitten edge, tips missing......	Mammals (p. 77)
V-shaped leaf bite, tips often lying on soil..	Birds (p. 75)
vi. Leaves torn, with ragged ends............	Slugs (p. 166) Leatherjackets (p. 162)
d. *Plant damaged after tillering stage*	
i. Whole plant sickly, yellow or reddish	
Root system short, much branched, white cysts visible June onwards.......	Cereal cyst nematode (p. 167)
Root system with small thickened galls.	Cereal root-knot nematode (p. 167)
Root system normal..........................	Grass and cereal mite (p. 165)
ii. Some shoots yellow, others healthy	Grass moth caterpillars (p. 49) Common rustic moth caterpillars (p. 54)
iii. Some shoots swollen, others healthy	Gout fly larvae (p. 157)
iv. Stem surface near upper nodes pitted with saddle-shaped depressions	Saddle gall midge larvae (p. 161)
v. Leaves bitten or discoloured	
Long narrow strips eaten from leaf blade	Slugs (p. 166) Cereal leaf beetle (p. 219) Barley flea beetle (p. 164)
Irregular areas bitten from leaf edge	Grass moth caterpillars (p. 49) Leaf sawfly larvae (p. 165)
Discoloured blotches on leaf surface	Aphids (p. 165)
Silvery marks on leaf surface	Thrips (p. 165)
Leaves with blister mines..................	Cereal leaf miner larvae (p. 161)

vi. Shoots bend or break and fall over

Shoot bent above ground level, brown seed-like bodies beneath leaf sheath near cut edge	Hessian fly larvae (p. 161)
Shoot cut off cleanly just above ground level	Wheat stem sawfly larvae (p. 165)
Shoot cut off with diagonal bite or peck marks, head often stripped	Birds (p. 75) Mammals (p. 77)
Shoot breaks off near upper nodes	Saddle gall midge larvae (p. 161)

e. *Damage to flowering heads*

Distal florets white, distorted or shrivelled .	Frit fly larvae (p. 160)
Whole ear white or aborted	Grass and cereal mite (p. 165) Hessian fly larvae (p. 161) Wheat steam sawfly larvae (p. 165)
Ears distorted, stems often twisted	Frit fly larvae (p. 160)
Ears often one-sided, groove runs down side of stem	Gout fly larvae (p. 157)

e. Ears sticky, often covered with extruded fluid .... Aphids (p. 165)

f. *Damage to ripening grains*

Grains missing	Lemon wheat blossom midge (p. 161) Birds (p.75)
Grain shrivelled or incompletely developed	Grain aphid (p. 165) Orange wheat blossom midge (p. 161) Thrips (p. 165) Birds (p. 75)
Grain blackened and/or partly eaten	Rustic shoulder knot moth caterpillar (p. 54) Frit fly larvae (p. 160)

Grain replaced by black powdery mass	Frit fly larvae (p. 160)
Grain flattened, white deposit on outside ...	Birds (p. 75)

A similar range of pests attacks barley and wheat in this country, with fewer species occurring on barley and this crop being more tolerant of pest damage. Close watch on fields growing spring barley either continuously or intensively has revealed little or no threat from pest attack. In wetter parts of the country, spring barley often follows grass or cereal stubbles infested with grass weeds, and under these circumstances leatherjackets and other 'ley pests' can cause serious losses.

INSECTS

Wheat bulb fly (see page 36)
Although winter barley is attacked when the newly-hatched grubs enter shoots from mid-January onwards, it is not a favoured host and larvae rarely survive to maturity within the barley tillers.

Spring barley is attacked occasionally, chiefly when fields badly infested with couch grass containing the maggots are ploughed and sown to spring barley within a few weeks. Recently, serious attacks have been reported when larvae damaged shoots below ground in spring sown crops following sugar beet.

Late wheat shoot fly (see page 42)
Winter and spring barley are very rarely affected.

Bean seed flies (see page 42)

Grass and cereal flies (see page 43)
Although winter and spring barley can be attacked, these crops rarely follow grass in the rotation and damage is noted less frequently than in wheat.

Yellow cereal fly (*Opomyza florum*) (see page 43)
Early-sown winter barley occasionally suffers damage from this pest, but not as badly or frequently as winter wheat.

Gout fly (*Chlorops pumilionis*) (Plates 49, 50, 51)
Attacks by gout fly on barley are often widespread in the southern half of England, but damage is much less severe than it was before 1950.

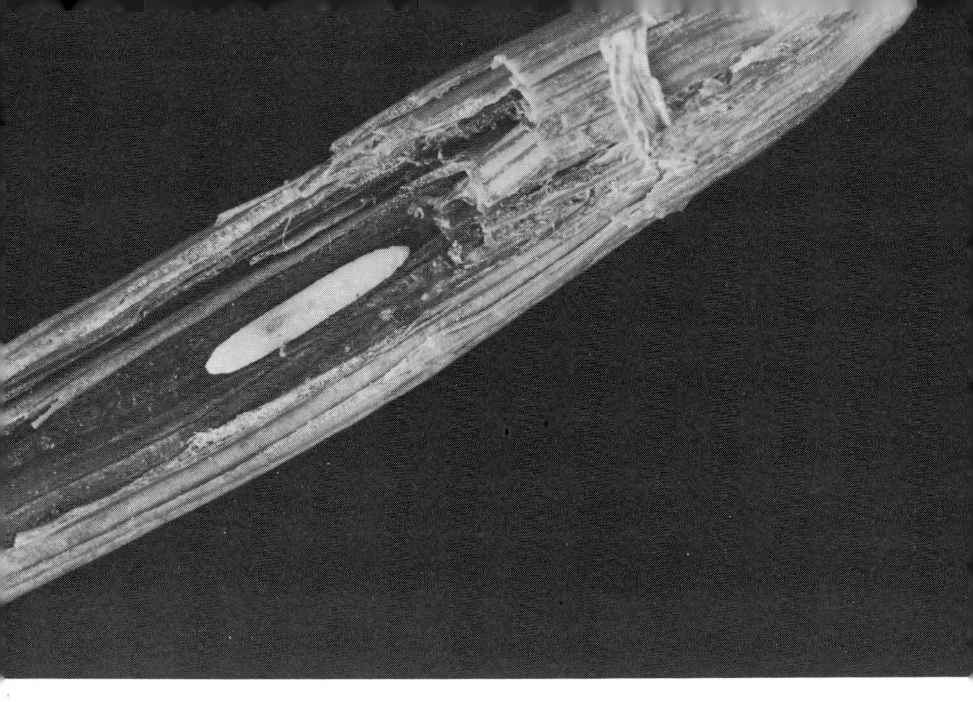

Plate 49. Gout fly. Cereal stem dissected to show larva (×2).

Plate 50. Gout fly. Barley plant showing effects of larval feeding. The left hand tiller is normal.

Plate 51. Gout fly. Barley ears damaged by larval feeding. Crown copyright.

Damage symptoms

The severity of damage varies according to the stage of plant growth. Very young barley seedlings may be killed. At tillering, plants may be stunted, and infested shoots have a swollen, gouty appearance, at first green in colour but later turning yellow with ragged leaf tips. Winter barley shows these symptoms from November until spring, whereas late-sown spring barley is gouty in June and July. Another kind of damage to spring barley occurs when the shoot lengthens and its terminal portion swells slightly; the contained ear either fails to emerge or is small and distorted. Spring barley sown in March or early April may be attacked just as the ear is about to emerge from the leaf sheath. Part of the ear may then be devoured and a groove runs down the stem from the damaged part of the ear to the flag leaf. Shoots in this instance look quite normal.

In each case described above, slitting the damaged plant will usually reveal the gout fly grub within as a creamy white, legless maggot about 1·8 cm long when fully grown.

Life history

The gout fly completes two and sometimes three generations each year. Adult flies first appear in May or June and are small and black and yellow. Females lay eggs, usually one per leaf, on the upper leaf surfaces of barley, wheat, rye and some grasses including couch grass. Oats and maize are immune.

The eggs hatch after a week or so and the maggot bores into the centre of a shoot, where it feeds for about a month and then pupates as a hard brown puparium. The pupal stage lasts a month and gives rise to adult flies of the autumn generation. These lay eggs in September on early-sown autumn cereals, volunteer cereal plants or grass weeds. The maggots which hatch from these eggs enter the shoots but take several months to reach maturity and do not pupate until March or April of the following year. These grubs are unable to migrate from ploughed-in cereals or grasses to a germinating cereal crop.

Control

Winter barley should not be sown too early, in order to avoid egg-laying in September and early October, or too late, thus avoiding damage by the spring generation. Spring barley should be sown before the end of April, for only May-sown or very backward crops are at risk.

In all cases, the seedbed should be adequately supplied with fertiliser, especially phosphate. Grass weeds should be removed because they can harbour the maggots. Chemical measures are not needed against gout fly.

Frit fly (see page 216)

Winter barley may be attacked in late autumn, the larvae entering the central shoots and giving rise to the 'deadheart' condition. White, thin maggots about 0·3 cm long or the brown puparia can be found within attacked shoots. In late spring, ears of barley may be damaged and the tips left white and empty. The presence of larvae or puparia is usually required to confirm that frit fly is responsible.

Leaf miners (see page 45)
Blotchy leaf mines are produced by larval feeding activities in much the same manner as in wheat. All barley cultivars seem to be susceptible.

Wheat blossom midges (see page 46)
Although barley acts as a host of both the lemon and orange blossom midges, damage is much less frequent than in wheat. Grains are shrivelled or aborted by larval feeding, and large differences in susceptibility of barley cultivars have been noted in other countries.

Saddle gall midge (see page 47)
Spring barley is equally prone to attack as wheat and yield losses recorded in eastern England in 1968–71 were greater in infested barley crops than in neighbouring wheat crops.

Hessian fly
Although the Hessian fly, *Mayetiola destructor*, has been frequently recorded throughout England since 1886 and in recent years in north-east Scotland, it has never reached the important status it holds in other parts of the world, especially North America.

Damage in this country is caused in early summer when individual stems of winter wheat or spring barley lodge in a manner similar to the 'straggling' due to eyespot. The whitish larvae of the midge or the brown ensheathed larvae ('flax seeds') can always be found near the point of stem breakage—just above the first, second or third node. In some cases larval feeding may not cause the cereal stem to break, but 'whiteheads' and poor quality grain may result.

Hessian fly has two generations each year in southern England but only one in north-east Scotland. In both areas it is the spring generation which causes damage. The adult midges emerge in May and June and lay eggs on wheat, barley, rye and couch grass. Oats and maize are not acceptable hosts. The eggs hatch in a few days and the maggots crawl beneath the leaf sheaths to feed just above a node. Larvae overwinter each within a brown skin ('flax seed') and may remain in this condition for up to four years. Pupation takes place in late spring.

Control measures are not advocated in this country and natural enemies help to keep the pest within acceptable limits. The recent upsurge of attacks on spring barley in eastern counties of Scotland

Plate 52. Leatherjackets. Damage to barley plants, with grubs exposed along the drill row. Crown copyright.

has usually occurred when the barley crop is sown close to one which was undersown the previous year.

In North America, the autumn generation of larvae can destroy winter wheat seedlings and much work has been done there on timing of drilling and on breeding of resistant cultivars of wheat.

Bibionid flies (see page 48)

Leatherjackets (Plate 52)
These are the larvae of crane flies (daddy longlegs) which are so familiar in grassy habitats in autumn. The adult fly is harmless but the grub is an important pest of cereals, grass and many other crops.

Damage symptoms
Barley seeds can be hollowed out in a similar fashion to slug damage. More usually leatherjackets attack established plants, biting off the stems at or just below ground level. The cut ends of the plant tissues are torn with frayed edges (see also wireworm

damage, page 55). Leaf tips may be eaten or even pulled below the soil surface.

Winter barley is attacked during the late autumn and winter months at a slow rate, but damage quickly increases in March. Spring barley is thinned from April onwards.

There is usually little difficulty in finding the greyish-brown, tough-skinned maggots, devoid of legs or an obvious head end, lying near damaged plants on or near the soil surface or sheltering under grass clods, often in the damper areas of a field.

As attacks frequently follow grass leys, leatherjackets are often included among 'ley pests'. Numbers may also be high in barley after cereals, in which case grass weeds in the preceding cereal stubble probably attracted the egg-laying female flies.

Life history

There are several kinds of crane flies whose larvae attack barley and other cereals, and it is difficult to separate them into species. For practical purposes it is sufficient to describe *Tipula paludosa*, for this is the most common tipulid fly whose larvae are responsible for most damage to cereals.

The adult *T. paludosa* has a brown body, one pair of narrow wings and long, spindly legs. The female fly has a pointed end to the abdomen, while the male has a truncate or swollen rear end. Both sexes are on the wing in autumn, and males are particularly attracted to light sources in September evenings. Eggs are laid in grassy places and the larvae hatch within ten days or so to feed on plant tissues or decaying vegetable matter through the winter months. Feeding rates increase in spring as leatherjackets near maturity and attain a length of 5 cm. Feeding ceases in June and the larva pupates just below ground. Just before adult emergence, the pupa pushes itself partly out of the soil.

Control

Young leatherjackets are extremely sensitive to drought conditions, and catastrophic losses in numbers may occur if the amount of rainfall during September and October is much below average. Virus diseases take their toll of the larvae, as do birds such as rooks, gulls and pheasants. Years of abundance, when damage to crops is rife in all but easternmost counties of England, are followed by years of scarcity when the only attacks are by relict populations in the wetter grassland areas of the south-west, Wales and northern England.

Ploughing grass in late July or early August should eliminate much risk of damage to the following cereal crop. Cereal stubbles infested with grass weeds should be cultivated or treated with an appropriate herbicide immediately after harvest.

As damage intensifies in spring, chemical control measures may have to be used. These involve surface poison baits, pellets or insecticidal sprays, the latter often preferred because of ease of application.

Baits of fenitrothion or gamma-HCH are incorporated with 30 kg/ha bran or are available as ready-made pelleted baits of gamma-HCH. Sprays of chlorpyrifos, gamma-HCH, quinalphos or triazophos should be applied in at least 225 l/ha.

Because of the risk of taint, root crops intended for human consumption should not be grown for eighteen months on soil following treatment with gamma-HCH. DDT should not be used. Chlorpyrifos does not work well in peaty soils. Rolling and a nitrogenous top-dressing help stricken crops to recover.

In epidemic years, leatherjacket numbers may reach several million per hectare and they can wipe out a cereal crop. Populations of more than 500,000 per hectare are usually damaging to spring cereals, although the pattern of damage should affect any decision on whether or not to apply chemical control measures. Widespread thinning of spring barley is of less consequence than complete loss of plant in discrete patches. As a general guide, if fifteen or more leatherjackets are found in ten 30 cm lengths of drill, chemical control measures are likely to be cost-effective.

Moth caterpillars

Those caterpillars which feed on wheat (page 49) are equally capable of attacking barley. Control measures, if required, are the same as those suggested for winter or spring wheat.

Beetles

Wireworms (page 55), chafer beetle larvae (page 57), wheat shoot beetle larvae (page 58) and cereal leaf beetle adults and larvae attack both winter and spring barley. Symptoms of damage are the same as for wheat and control measures are identical.

Barley flea beetle

Phyllotreta vittula attacks not only barley but also wheat, oats, rye, maize, grasses and brassica crops. The adult overwinters in hedge bottoms and other sheltered places, then emerges to feed on host plants, cutting small narrow strips from the interveinal leaf tissues. The beetle is only 0·24 cm long and is black with two wide yellow stripes running along the back. Only one generation is completed each year and damage to cereals rarely if ever needs control measures.

Sawflies
Wheat stem sawfly (page 59) occasionally damages spring barley, the symptoms resembling those in wheat.

Other sawfly larvae, notably of *Dolerus* and *Pachynematus*, are commonly found in summer nibbling the edges of leaves or stretched along the barley awns. The damage is of no consequence and control measures are unnecessary.

Aphids and leafhoppers
Those aphids found on wheat (page 60) also infest winter and spring barley. Damage symptoms from direct injury are less obvious than in wheat or oats but some aphid species are important as vectors of barley yellow dwarf virus (page 200).

Although infestations of *Sitobion avenae* and *Metopolophium dirhodum* have been shown to reduce grain yield by 375 kg/ha, both winter and spring barley cultivars are more tolerant of direct aphid injury than wheat or oats. Significant yield losses are recorded only when over 50 per cent of the plants are heavily infested, and in recent trials in Essex and Kent the yield response of sprayed compared with unsprayed plots did not usually justify treatment with an aphicide.

Further work is needed on the effects of leaf-feeding aphids on winter and spring barley.

Some progress has been made in breeding for resistance to barley yellow dwarf virus, as in the winter barley cultivar Vixen.

The leafhopper *Javesella pellucida* (page 60) is found on barley and can transmit the virus causing wheat striate mosaic disease.

Thrips (see page 65)
Barley is less frequently damaged than wheat, oats or maize (but see comments on page 65).

MITES (see page 65)

The grass and cereal mite, *Siteroptes graminum*, is sometimes found on barley but it is economically unimportant.

MILLEPEDES (see page 66)

SLUGS (see page 68)

Attacks on winter and spring barley occur much less frequently than on wheat since the barley seed is protected by the palea and lemma. Early damage to germinating barley grains is sometimes confused with that caused by leatherjackets. However, serious attacks have been noted to barley seedlings before tillering. Control measures are described in the section dealing with wheat pests (page 70).

NEMATODES (EELWORMS)

In general, barley tolerates attacks by nematodes more readily than either wheat or oats.

Root-ectoparastic nematodes (see page 73)
Trichodorus, *Paratrichodorus* and *Longidorus* nematodes affect barley growwth and are commonly found in light soils in which spring barley is the dominant cereal. A relationship between initial numbers of *L. elongatus* and yield of spring barley has been established in trials in the West Midlands and in the Vale of York.

Other species of *Longidorus* known to damage barley in England are *L. leptocephalus* and *L. vineacola*. Stubby-root nematodes (*Trichodorus, Paratrichodorus* spp.) are involved in the stunting of winter and spring barley in sandy soils in Yorkshire and elsewhere. At densities of about 300 per litre of soil, *P. anemones* can adversely affect yield of spring barley. Yield responses of up to 1·25 t/ha have been recorded following granular nematicide soil treatment.

Stunt nematodes (*Tylenchorhynchus* spp.), sheath nematodes (*Hemicycliophora* spp.) and pin nematodes (*Paratylenchus* spp.) are commonly found in British soils. Their effects upon the growth and yield of barley are not known but they must rank as pests of only minor importance.

Root-lesion nematodes (see page 74)
Species of *Pratylenchus* are often found in large numbers within barley root tissues, feeding on the cortical cells and causing cavities to develop within the roots. These nematodes do not generally have a significant effect on grain yield. *P. fallax* is very pathogenic to wheat, barley and maize roots and is exceptional in being associated with patches of stunted spring barley plants grown on the Bunter sandstone of Nottinghamshire. Yield responses of up

to 21 per cent have followed soil fumigant use. It has not been found possible to establish a clear relationship between initial numbers of *P. fallax* in the soil and grain yield.

Cereal root-knot nematode (see also page 74)
Meloidogyne naasi is widespread in Wales, southern England, many European countries and North and South America. In Wales in 1977, 75 per cent of barley fields examined contained galled plants.

Affected crops show areas of poor growth from late spring onwards. Individual plants are stunted with older leaves yellowed and desiccated from the tip downwards. Infested plants produce fewer tillers and later flowering heads with smaller grains than healthy plants. These symptoms are often masked when infested areas are associated with poor soil conditions. Examination of the root systems of infested barley plants in June will reveal the swollen galls containing the developing larvae.

All cultivars of winter and spring barley so far tested in this country have proved to be efficient hosts of *M. naasi*, although some cultivars are more prone to damage than others. In the United States of America, four races or pathotypes of the nematode have been identified, making any resistance breeding programme complex.

Field tests in Wales have shown that yield losses with spring barley are 3·5 per cent per ten larvae per gram of soil, with 50 per cent loss of yield at an initial level of 150 larvae per gram. In many cases in the field, adverse soil conditions such as waterlogging may be present and then contribute towards poor yields.

The only practical means of control at present lies in crop rotation, using oats, potatoes and brassica crops to reduce nematode numbers in the soil. Oilseed rape varieties on the NIAB Recommended List are sufficiently poor hosts to qualify for use as break crops. It is likely that *M. naasi* has increased in significance in Wales in recent years because barley has supplanted oats on many mixed farms.

Cereal cyst nematode (see page 221)
Cysts of this nematode (*Heterodera avenae*) can often be found on the roots of barley; the pest is therefore regarded as a potential threat to continuous or intensive cropping with spring barley. Although barley is an efficient host, the pest fails to multiply on that cereal in many seasons and its numbers fluctuate at low, non-damaging levels. In addition, fungal parasitism of the cysts has

reduced the populations of this nematode and serious attacks are now rare.

That *H. avenae* can affect spring barley yields has been clearly demonstrated in recent years. Losses of 75 kg/ha in Pallas barley have been recorded for every increase of ten eggs per gram of soil. Comparisons of resistant and susceptible barleys, differing only in the presence or absence of the genetic factor controlling resistance, have given yield increments of as much as 20 per cent when both were grown in infested soil. The yield advantages of resistant barleys vary with initial nematode populations and multiplication rates. Trials using soil nematicides have also given increases of 1000 kg/ha in Proctor barley, although the nematicides were applied at uneconomic rates and were affecting other soil organisms in addition to *H. avenae*.

We thus have a situation in which the nematode is widespread and particularly common in those chalkland soils of southern England where spring barley is grown intensively. In most years, yield losses are very slight and the pest does not increase greatly in numbers. Although barley root systems are affected similarly to those of wheat and oats, the barley plants above ground shows little or no sign of being infested.

Control

Chemical control of *H. avenae* for barley crops is at present ruled out on economic grounds. Cultural control entails growing non-host crops for at least two successive years on the same field, when nematode numbers can be expected to fall considerably. For those who farm some of our most heavily infested land, break crops such as long-term grass, oilseed rape or field beans are simply not practical alternatives to spring barley. Hope lies chiefly in the breeding of resistant barley cultivars, which yield well in infested soil and give a greater decrease in post-cropping nematode numbers than either a fallow or break crop. Resistance from barley No. 191 has been incorporated into cultivars adapted to British agriculture and is currently available in the spring barleys Sabarlis and Tyra. They contain the single dominant gene (Rha 2) which confers resistance to the two races (pathotypes) of the nematode widespread in the United Kingdom.

BIRDS (see page 75)

MAMMALS (see page 77)

Chapter 5

BARLEY DISEASES

PLANT DAMAGE SYMPTOMS

Disorder	*Cause*	*Symptom*	*Page*
SEEDLINGS			
Failure to emerge	Organomercury injury	Seed does not germinate or seedlings with short thickened shoots, stunted roots.	23
	Gamma-HCH injury	Similar to and often associated with organomercury injury but roots and shoots are club-shaped.	23
Delayed emergence	Deep drilling	Seedlings yellowish, delayed sometimes in patches.	84
	Blue mould	Blue mould on seed grain.	84
Seedling blight	*Fusarium* spp. *Pyrenophora graminea*	Seedlings killed before or just after emergence.	176 188
Browning root rot	*Pythium* spp.	Patches of yellowed seedlings, roots brown.	173
Stunting and root rot	*Rhizoctonia solani*	Poor growth in distinct patches.	173
Snow mould	*Fusarium nivale*	Patches yellow, plants with pink fungal mould especially after snow.	176
Snot rot	*Typhula incarnata*	After snow patches of yellowish or rotted plants; small round brown fungal bodies on dead plant tissues.	174
Frost lift	Winter frost	Small or large patches of sickly plants in spring.	86
Leaf tipping	Frost damage	White leaf tips and	210

Disorder	*Cause*	*Symptom*	*Page*
		spots on leaves of young seedlings	
May yellows	Temporary nitrogen deficiency	Extensive areas, often whole fields, show leaf yellowing.	175

ROOTS AND STEM BASES

Disorder	*Cause*	*Symptom*	*Page*
Take-all	*Gaeumannomyces graminis*	Roots, and later some stem bases, blackened; in patches or at random; plants prematurely ripe, sometimes stunted.	175
Brown foot rot	*Fusarium* spp.	Stem bases brown, later with pinkish spore masses; stems break easily at node just above soil level; mainly at random.	176
Eyespot	*Pseudocercosporella herpotrichoides*	Eyespot lesions on stem bases, often girdling stem; can cause lodging especially in winter barley.	176
Sharp eyespot	*Rhizoctonia cerealis*	Lesions similar to but more clearly defined than for eyespot. Later sometimes large 'watermarks' on sheath.	177
Foot rot	*Cochliobolus sativus*	Rot at stem base, uncommon.	177

LEAVES AND STEMS

Disorder	*Cause*	*Symptom*	*Page*
Yellow rust	*Puccinia striiformis*	Orange-yellow pustules arranged in lines.	177
Brown rust	*Puccinia hordei*	Brown pustules scattered at random on leaf, becomes epidemic late in season.	179
Black stem rust	*Puccinia graminis*	Orange-brown and later black pustules mainly on stem, rare.	181
Mildew	*Erysiphe graminis*	White fluffy pustules, on leaf surface.	181

Disorder	*Cause*	*Symptom*	*Page*
Leaf stripe	*Pyrenophora graminea*	Long brown stripes on leaves; successive leaves on a plant affected. Affected plants at random.	188
Net blotch	*Pyrenophora teres*	Short brown stripes, sometimes blotch of network of brown lines, at random on leaves. General in crop.	189
Leaf blotch	*Rhynchosporium secalis*	Fairly large pale brown or grey blotches with dark margins.	194
Halo spot	*Selenophoma donacis*	Small angular spots each with pale centre, dark margin; black spore cases just visible to naked eye in lines on centre of spot. Usually on flag or youngest leaves.	197
Leaf spot	*Septoria nodorum*	Brown blotches on leaf blade and sheath; sometimes black spore cases on centre of spot.	198
Other spots and blotches	Various	Small black-brown spots especially on older leaves.	199
Barley yellow dwarf	Barley yellow dwarf virus	Leaf bright yellow from tip extending to base; sometimes only upper leaves (especially flag leaf) affected; in spring barley usually isolated plants at random but often patches in winter barley.	200
Barley yellow mosaic	Barley yellow mosaic virus	In winter barley, distinct patches small or large in Feb–April. Plants spiky with leaf mottle.	205
Manganese deficiency		Leaves pale green with very small pale brown spots; floppy appearance; mainly in patches at late tillering.	207

Disorder	*Cause*	*Symptom*	*Page*
EARS			
Covered smut	*Ustilago hordei*	Inside of grain replaced by mass of black spores held in place by transparent skin; spores not shed; uncommon.	207
Loose smut	*Ustilago nuda*	Grains replaced by mass of black spores, obvious at emergence but spores are then shed leaving bare stem.	208
Rusts	*Puccinia* spp.	Yellow-brown pustules on awns and glumes, usually brown rust but see leaf symptoms.	177 179
Ergot	*Claviceps purpurea*	Hard black bodies replace grains in a few spikelets.	209
Scab	*Gibberella zeae*	Red fungus with black spore cases embedded; on surface of ear.	144
Ear blight	*Fusarium* spp.	Pink-red fungus on surface of ear.	97
Botrytis	*Botrytis cinerea*	Individual spikelets at random; grain destroyed, grey mould usually visible.	209
Black mould	*Cladosporium* spp.	'Sooty' black moulds on surface; ears often thin.	210
'Whiteheads'	*Gaeumannomyces graminis, Pseudocercosporella herpotrichoides, Fusarium* spp. Frost damage Herbicide injury, etc.	Ears bleached, prematurely ripe; grain shrivelled; a symptom of many disorders (see also root and stem base diseases).	
'Blindness'	Frost damage	Many or few spikelets bleached and blind in groups on one part of the ear, frequently at the tip (many causes of 'blindness' other than frost.).	210

Disorder	*Cause*	*Symptom*	*Page*
'Blindness'	Uncertain	Spikelets empty not bleached.	211
Copper deficiency		Terminal-spikelets blind; ears rat-tailed; grain shrivelled.	211
Distortion	Herbicide injury	Ears malformed.	211

Browning root rot (*Pythium* spp.)
Barley is susceptible to this disease but very few records exist of its occurrence on the crop in Britain. This is almost certainly because it is not easily recognised, producing only a general browning of roots which is a fairly common condition even in apparently healthy barley which yields well. It is possible that this root rot, along with some others as yet unidentified, may be responsible at least in part for the plateau in average barley yields in intensive systems but this remains to be examined in more detail. The disease is discussed under wheat (page 85).

Barley stunt disorder (*Rhizocotonia solani*)
Cases of stunting occurring in patches especially in barley are reported each year from widely distributed districts but almost invariably from crops grown on light sandy soils. The cause of the disorder was uncertain for many years but it is now generally agreed that the principal cause is the soil-borne fungus *Rhizoctonia solani*, although other factors may also be involved.

Both autumn- and spring-sown barley crops can be affected. A similar disease occurs less commonly in wheat (page 85) probably because this crop is less frequently associated with light soils.

The disease usually occurs in small (1–3m) well-defined roughly circular patches but sometimes more extensive areas, often kite-shaped, are affected. Typically about 1–10 per cent of the total area of the crop is affected but occasionally this may extend up to 50 per cent. Usually little or no grain is harvested from affected patches.

Symptoms are seen, soon after plants emerge, as leaf yellowing and stunting; young plants may develop a purplish discolouration (at one time the condition was described as 'purple patch').

The causal fungus mainly infects the roots causing browning and rotting; it can sometimes be found in the coleoptiles and leaves of affected plants. The outer cortical layers of the root are rotted first and as patches of this tissue are lost the roots assume a beaded appearance and the tips of the roots are often finely pointed where only the vascular tissue remains. This damage to the root system is often accompanied by a considerable proliferation of lateral roots.

Some plants may be killed at an early stage but others survive and recover as the season progresses, though patches remain clearly discernible. Affected plants tend to remain green longer but in spite of some recovery rarely produce harvestable grain. The disease may not be noticed until the later stages when it is more difficult to isolate and identify the causal organism.

Barley stunt disorder is often worse after ploughed-up grass and affected patches tend to recur in the same place for three or more years in successive crops of barley. The disease also affects other arable crops, notably sugar beet, in the same patches.

There are no practicable control measures.

Snow rot (*Typhula incarnata*)

This disease was uncommon in the United Kingdom but was found more frequently as the area devoted to winter barley increased. The disease occurs in all areas but is most common in Scotland and in other places where there is snow cover in winter. The disease is noticed when the snow melts away and is favoured by lengthy periods of snow cover. The disease should be distinguished from 'snow mould' caused by *Fusarium nivale* (page 176).

Snow rot usually occurs in patches which are noticeably pale brown. Leaves and stems are affected and plants may be killed, resulting in a thinning of the stand or bare patches. The bases of affected plants bear a white mould with numerous and characteristic reddish-brown fungal bodies (sclerotia) which are about the size of clover seed. These sclerotia survive in the soil and germinate in autumn to infect the new cereal crop directly or to produce basidiospores, an air-borne source of infection.

In Scotland the disease has been severe in crops sown early and when mild autumn weather has favoured a lush growth. Such conditions are ideal for rapid spread from plant to plant within and between rows under the snow cover. In thinner crops from later sowings the disease tends to be less severe with spread restricted to within the rows.

Varieties vary significantly in their reaction to the disease, from moderately resistant to very susceptible.

In most areas the disease thins out the crop but in the spring there is considerable recovery and the disease is not usually damaging. In some areas where damage occurs in most years, it may be necessary to apply fungicide treatments in the autumn. A good control can be obtained from some of the triazole fungicides (page 26) which, as seed treatments or as sprays, also give control of some other leaf diseases and snow mould (page 76). Sprays of benodanil are also effective but do not control other diseases likely to be present in the autumn. For the best results the spray should be applied as late as possible but before snowfall. Such a timing presents serious difficulties and partly to overcome these, two treatments may be recommended: a seed treatment or an early spray followed by a spray applied as late as possible.

May yellows
This term describes a condition in spring barley crops in May when extensive areas seem to stop growing and the green leaves turn yellow. It usually occurs immediately after periods of cool weather and heavy rains and is associated with a temporary shortage of soil nitrogen due to severe leaching and reduced activity both of the soil nitrifying organisms and of the barley root system. The symptoms show when plants begin to grow again on the return of warmer weather. With a spell of favourable growing weather the symptoms soon disappear and usually there is no need to apply extra nitrogen.

Under similar conditions magnesium deficiency symptoms may occur. These show as pale green areas in bands between leaf veins. The trouble is transient and the crop does not need treatment except on certain soil types such as some black fen soils.

A similar range of symptoms may be seen in other cereals especially in wheat.

Take-all (*Gaeumannomyces graminis*, syn. *Ophiobolus graminis*)
The symptoms of take-all in barley are similar to those in wheat (page 87). However, barley is inherently more tolerant of the disease and damaging attacks are seen less frequently. Occasionally the most severe attacks occur in crops of winter barley sown very early (August–early September) when the disease shows in the late autumn in small patches of pale green or yellow plants with severely diseased roots.

Much of the barley crop is sown in the spring and under modern farming conditions this is rarely damaged significantly by take-all. Not only is barley inherently more tolerant but the spring-sown crop is subjected initially to lower levels of inoculum because of the longer interval between the previous harvest and sowing. The spring crop is in the ground for substantially less time than the autumn-sown crops and additionally it benefits from the more favourable growing conditions for most of its life. All these factors combined with good husbandry practices have enabled many farmers to practice successfully the growing of spring barley intensively and even continuously.

Many farmers have also been able to practise growing winter barley intensively or continuously though there are clearly more risks with this crop than with spring barley.

Intensive or continuous barley growing can be sustained provided the physical conditions of the soil are good and a high standard of husbandry is practised. It is essential to obtain good seed bed conditions so the crop is well established and to maintain the control of weeds especially grass weeds which may carry the take-all fungus. These systems are also dependent upon high levels of nitrogen application for satisfactory yields.

Factors affecting take-all and its severity are discussed under wheat (page 87) and the general principles also apply to barley, especially winter barley.

Seedling blight, snow mould, brown foot rot, ear blight and **scab** (*Fusarium* spp.)

Barley is considered to be comparable to wheat in its susceptibility to this group of diseases and the account of the disease in wheat (page 95) also generally applies to barley, especially winter barley.

Fusarium nivale causes snow mould which together with snow rot (page 174) can cause extensive damage to winter barley under a snow cover. The affected plants have yellowed leaves often covered by a pink fungal mould.

Scab (*Gibberella zeae*) (see page 144) has been recorded occasionally on barley in Britain, usually in western or northnorthern parts.

Eyespot (*Pseudocercosporella herpotrichoides* syn. *Cercosporella herpotrichoides*)

Although eyespot is best known as a disease of wheat (for a full account see page 98), barley is also affected. Spring-sown barley, like spring wheat, escapes the worst effects of the disease except for early-sown crops in the wetter areas of the country. More attention has been paid to the disease on winter barley since the area sown to this crop in the United Kingdom has increased. Field experience indicates that varieties currently grown exhibit resistance at least as effective as that available in the most resistant winter wheats.

In the early growth stages in the spring before stem elongation, other diseases may produce lesions on the leaf sheaths of winter barley which are superficially similar to the early stage of eyespot, making diagnosis difficult. These other diseases tend to be confined to the outer leaf sheath but where significant eyespot infection occurs the fungus penetrates the underlying tissues and lesions can then be found on some of the inner sheaths.

Whilst resistance can be relied upon to give adequate control in most circumstances, in conditions especially favouring infection (early autumn sowing, lush growth, mild wet winter, heavy soils) winter barley can be severely affected and control with the same fungicide as used on wheat is cost effective. However, because other diseases commonly affect barley during the late tillering/ early jointing stages it is often most cost effective to use a broad spectrum fungicide or compatible mixture that is effective against all diseases (mildew, rusts, net blotch and rhynchosporium) as well as eyespot.

Sharp eyespot (*Rhizoctonia cerealis*)
Sharp eyespot is fairly common in barley especially winter barley, but it appears to be less of a problem than it is in wheat. The account of the disease on wheat (page 106) also largely applies to barley.

Foot rot (*Cochliobulus sativus* [*Drechslera sorokiniana* syn. *Helminthosporium sativum*])
This disease may be more common than had been supposed at one time though it rarely causes significant yield losses. It attracted attention in the 1970s when a very susceptible variety, Clermont, was grown. A loss in yield of about 15 per cent was recorded when a stock of this variety with extensive seed infection was grown.

The disease causes brown lesions at the base of the leaf sheaths near soil level and the disease may progress to cause a foot rot involving stem bases and roots. The disease also causes a leaf spot. It is seed-borne and this phase is not completely controlled by organomercury treatment. The fungus can survive over winter in the soil and on other graminaceous hosts.

Yellow rust (*Puccinia striiformis*)
This is an important disease of wheat (see page 109 for a detailed account) and while it also occurs on barley it has been a much less important disease nationally on this crop. Changes to a much higher proportion of the barley crop sown in the autumn initially

resulted in an increased incidence and severity of yellow rust on both winter and spring barley. However, this was not sustained, possibly because more resistant varieties were grown.

The fungus, which is an obligate parasite growing and surviving only on green plants, is largely or entirely specialised to barley. The symptoms consist of lemon-yellow pustules arranged in lines on the leaf blade and leaf sheath. Although the lines of pustules are usually grouped to form stripes on the leaf blade, they can also occur in patches on the leaf especially on young leaves in the early growth stages; the ear can also be affected. Later in the season black pustules are found scattered amongst the yellow ones. Yellow rust may be distinguished from the more common brown rust by the colour and arrangement of the pustules on the leaf blade.

Yellow rust of barley is not uncommon but it has rarely become epidemic or caused severe damage. However, the disease has a potential for causing severe damage and may reduce yields by up to 50 per cent. This was illustrated by the severe outbreaks in the mid 1970s in some spring barley crops in the north of England. These severe attacks were associated with the growing of susceptible winter barley varieties which, in some mild winters, carried large amounts of inoculum into the spring. The disease then spread in early June to spring barley crops and some very susceptible varieties which were then being grown were severely damaged. The relative infrequency of severe attacks of barley yellow rust in the past was probably largely due to the use, mainly unknowingly, of varieties with good levels of general resistance. The restricted area of winter barley crops carried only low amounts of inoculum through to the spring and the development of the disease on these and on the more common spring-sown crops was delayed, usually so late that little damage was caused. The yellow rust fungus requires leaf wetness to become established and is generally favoured by cool wet weather so that by the time the disease normally occurs in spring barley, epidemic spread would tend to be restricted by the warmer and drier summer weather.

Although the pathogen occurs as distinct races only a few have been recorded from barley in Britain. This is associated with the infrequent use of major gene (specific) resistance against yellow rust in barley by plant breeders. Good general (non-specific) resistance exists in both winter and spring varieties and, by avoiding highly susceptible varieties, it seems likely that good control of the disease can be maintained.

Effective fungicides are available for the control of barley yellow rust. On winter barley it is suggested that fungicides should be applied if the disease is noticed on the new spring growth; further applications may be needed on susceptible varieties. In spring barley, fungicides should be applied to very susceptible varieties as soon as the disease is noticed but not later than the flowering stage. Other more resistant varieties are unlikely to require fungicide treatment.

Brown rust (*Puccinia hordei*)
Brown rust is regarded as the second most important leaf disease of barley, after mildew. In most seasons it causes relatively small losses but in 1970 and 1971 it was particularly severe causing serious yield losses in many crops, especially in the southern half of Britain. This rust, which occurs only on barley, is distinct from brown rust of wheat. The cluster-cup (aecidial) stage occurs on species of *Ornithogalum* and in England has been found on *O. pyrenaicum*. Although the stage on *Ornithogalum* may provide the fungus with the mechanism for the production of new races, it is thought to be unimportant in the epidemiology of the disease in this country. Certainly the fungus does not depend on its alternative host in order to survive from season to season.

Symptoms
The orange-brown pustules containing uredospores are distributed at random on the leaf blade. They also occur on the sheath, where they may tend to occur in short lines along the veins, and on the ear where pustules are clearly seen, especially on the awns. Later in the season black pustules containing teliospores are produced, scattered among the orange-brown pustules. The pustules tend to be small and the ones formed in the later stages of a severe attack are particularly small and may be difficult to recognise on leaves killed by the disease.

Sources and spread
Brown rust is said to survive the harvest period on late green tillers. Field observations suggest that there may also be some limited carry over of the fungus as spores in the soil or on debris as has been reported for brown rust of wheat (page 114). The fungus spreads to volunteer barley and winter barley and survives the winter on these plants as pustules or invisibly as mycelium inside the leaves. In the spring the disease tends to spread slowly on both winter and spring barleys and often only becomes

conspicuous later in the season, usually after ear emergence. Moderately severe attacks occur on winter barley but much more severe attacks may be seen on spring barley a few weeks later. The reasons for severe attacks building up late in the season are imperfectly understood but the more favourable temperatures in late June and July are probably a major factor. Late-sown spring crops which are subjected to high levels of inoculum at an earlier stage of their development, when environmental conditions tend to favour the disease, can be very severely affected.

As in the case of mildew, winter barley is the most important source of brown rust for the spring barley crop and disease gradients, declining with distance from the winter barley, can be detected in the surrounding spring barley in June. The brown rust spores are wind-borne and by this means long-distance spread of the disease can occur, though this is much less likely to result in early epidemics than infections deriving from nearby sources.

Disease development

Infection and pustule development can take place over a wide range of temperatures, from as low as 3°C to as high as 30°C, but most rapid development of the disease occurs at 15–25°C with an optimum around 20°C. The time between infection and sporulation is six to twelve days in the summer but may be as long as thirty days during the winter and early spring, and this probably accounts for the slow development of the disease at those times of year. High humidity and probably free water, from rain or dew, are necessary for spore germinations and infection.

Effect on the crop

The estimated average annual yield loss caused by brown rust in spring barley (based on ADAS leaf disease surveys) for the period 1967–80 was 1·5 per cent. This includes estimates of 8 per cent in an epidemic year such as 1970 but ranges from 0·3 to 2·0 per cent in non-epidemic years. In experiments the control of severe brown rust with fungicides has given yield increases of up to 25 per cent. Since brown rust develops late in the season the main effect is on grain size and severe attacks result in much shrivelled grain. Severe attacks in late-sown crops result in many grains failing entirely to set and many more failing to mature. In contrast to mildew, brown rust can be severe on leaf sheaths and ears so that the disease continues to affect yield adversely after the leaf blade has been killed.

The severe epidemics of barley brown rust in 1970 and 1971

were associated with the cultivation of highly susceptible varieties on a large proportion of the area devoted to spring barley. Since then such varieties have been avoided and widespread severe attacks have not occurred although all barley varieties grown in Britain are susceptible to the disease. Plant breeders have mainly avoided the use of high levels of resistance based on specific major genes, but instead have exploited a form of partial resistance known as 'slow rusting' derived from several sources. Varieties with this type of resistance are susceptible but most stages in the development of the disease occur at a much reduced level so that spread and development in a crop occurs more slowly and the disease causes correspondingly less damage.

Control

The use of varieties with good resistance of the 'slow rusting' type should provide an adequate control of barley brown rust (see NIAB List of Recommended Varieties).

If the more susceptible varieties are grown they should preferably not occupy a large proportion of the barley area on any one farm, especially in the southern half of the country where the disease is most damaging. Susceptible spring varieties should also not be grown near to winter barley which is the most important source of inoculum for early infection. Late-sown spring barley is most likely to suffer severe losses, particularly if it is sown adjacent to winter barley or early-sown spring barley.

Effective fungicides are available for the control of barley brown rust but severe levels of the disease occur too infrequently and erratically for their routine use to be recommended. If the disease is noticed in the more susceptible varieties before the end of flowering then a fungicide should be applied. The presence of low levels of brown rust in winter barley in the spring is not uncommon and a spray should not be necessary at this time unless the variety is known to be very susceptible.

Black stem rust (*Puccinia graminis*)

The orange and, later, black pustules are found mainly on the stem. This disease is rare in barley (but see wheat page 116).

Mildew (*Erysiphe graminis*)

Mildew, more correctly called powdery mildew, is the most serious leaf disease of barley. From data collected in surveys of spring barley, in England and Wales in 1967–80 it was estimated that the average annual yield loss due to mildew was about 10 per cent and ranged in particular years from 4 to 14 per cent. Furthermore, it was estimated that this disease was responsible for more than half the losses caused by all cereal leaf diseases and this took into account the widespread use of barley mildew fungicides since the early 1970s.

Symptoms

The disease is easily identified by the white pustules which occur on the surface of the leaf blade and sheath and sometimes coalesce to cover the whole leaf. In barley a common symptom is small brown necrotic spots on which the white fungal mycelium may be obvious but sometimes can be seen only with the aid of a hand lens. Such symptoms are associated with resistance to infection due to genetic (varietal) or physiological factors. The spots may be few or many in number and the symptoms can be associated with significant yield losses.

The form of the fungus which affects barley is specialised to barley so that the only source of the fungus is barley. The symptoms and life history of the disease are similar to those of wheat mildew (for a detailed account see page 118). In contrast to wheat, mildew does not attack barley ears and the form of the epidemic on both autumn- and spring-sown crops is different from that on winter wheat.

Sources

The fungus is an obligate parasite growing only on living green plants. It survives the harvest period as black spore cases (cleistothecia) on the stubble and the remains of the previous barley crop. The ascospores released from the cleistothecia infect volunteer barley plants and spores (conidia) produced on these infect winter barley. Late green tillers which survive harvest are also a source of conidia. Early-sown winter barley may be infected by ascospores in addition to conidia from volunteers and, because temperatures are favourable from September to October, these crops may be severely attacked. Crops which emerge after late October escape severe infection but usually carry some mildew through the winter and are frequently the crops most affected by the disease in early spring. The main source of mildew for spring barley is conidia produced on winter barley. Volunteers which survive the winter in unploughed stubbles or in leys may also carry mildew through the winter, but they are thought to be a negligible source of the fungus compared with winter barley.

Spread

The disease may spread rapidly and cause severe damage to crops sown in September and which emerge during the warmer months in the autumn. Later the cold weather restricts spread though the disease may spread significantly during a mild winter. More rapid further spread occurs mainly as temperatures rise in the spring. Mildew develops within the winter barley first and then spreads to adjacent spring barley crops through air-borne conidia. In spring barley, disease gradients are usually detected only within a distance of about 200–400 metres of the winter barley, but they undoubtedly exist over greater distances. There is some evidence of long-distance spread of over several hundred miles by means of air-borne spores. Spread from adjacent crops is especially important in establishing early infection but once the disease occurs in a crop then conditions within the crop are most important in determining the rate at which the epidemic develops. Once the disease is established in spring barley crops, they act as additional, and in time, the principal, foci for the further spread of the disease.

Mildew is able to develop and spread under a very wide range of conditions; development is favoured by warm dry weather, with optimum temperature about 20°C. Spread occurs more slowly at lower temperatures and spore production is restricted above 25°C. Barley is most susceptible when the relative growth rate of the crop is highest. In winter barley this is in early autumn and early spring and, for most spring-sown crops, is in the period May–early June, depending on sowing date and the weather. If favourable weather occurs relatively late in spring and early summer, early-sown crops may have passed through their most susceptible phase and then escape severe attacks; this often occurs with autumn-sown barley. Crops sown in late spring on the other hand are most susceptible in June, when favourable weather is most likely to occur, and so frequently suffer severe attacks.

Effect on the crop

Experiments have shown that mildew can cause yield losses of 40 per cent or even more but usually losses are of the order of 5–20 per cent. The yield loss formula (per cent loss = $2{\cdot}5\sqrt{M}$ where M = the percentage leaf area affected by mildew at complete ear emergence) gives a useful guide to losses in spring barley. However, the formula tends to under-estimate losses when the attack is severe and especially at the very early growth stages. It is a poorer guide for winter barley, in which loss relationships depend upon the severity of attack at different times during the long growth period of the crop. Early attacks can reduce root growth severely in both winter- and spring-sown crops, rendering winter barley more susceptible to winter kill, especially on thin or light soils, and spring barley more susceptible to summer drought. Such attacks can also reduce the number of fertile tillers and the size of the ear. Severe attacks that develop after tiller number and ear size have been determined affect the size of the grain and to a lesser extent the number of grains per ear.

Control

Culture. The manipulation of cultural factors will not provide satisfactory control of mildew on its own but can have a significant effect upon the severity of the attack. In winter barley serious attacks in the autumn may be avoided by drilling so that the crop does not emerge until late October. However, this carries such a large penalty in terms of reduced yields that it is commercially unacceptable. Very early sowing, before mid-September, should be avoided. Late tillers and volunteers on which the fungus survives after harvest should be destroyed especially if winter barley is to be grown nearby. Since winter barley is the important source of mildew for spring barley, the sowing of spring varieties close to winter crops should be avoided. Where it is unavoidable, the very susceptible varieties should not be used and early attacks should be controlled by seed treatments or early fungicide spray.

Late-drilled spring barley crops are particularly susceptible to mildew, especially when near a source of mildew (autumn or earlier spring-sown crops) and should be avoided or receive prophylactic fungicidal treatment. Over-generous nitrogen fertilisation encourages the disease but less than the optimum treatment may involve yield losses as great or greater than those caused by the disease. Applications of potash or phosphate above the optimum for the crop are not likely to be beneficial.

Varietal resistance. The most economic method of control would seem to be the use of varieties with a very high degree of resistance amounting to near immunity. Such varieties have been available but their success has been short-lived. Resistance in these varieties was of the specific kind based on only one or two major genes and within a year or two of being widely grown, such varieties became susceptible to new races of the fungus. This often occurred when a substantial area was devoted to a particular variety and, since most of the varieties then proved to be very susceptible to mildew, they suffered severely and were soon discarded. In an attempt to utilise these sources of specific resistance more successfully the concept of diversification has recently been developed (page 21). Diversification in respect of resistance factors is applied to the selection of varieties. The object is to reduce the rate of spread by choosing varieties with different resistance factors to be grown in adjacent fields. An extension of this idea is to grow in one field a mixture of three or more varieties or lines, each containing a different source of resistance. This has been shown in experiments to reduce the spread of mildew to about half that expected. The level of control achieved by the use of mixtures is not always satisfactory on a field scale and fungicides may have to be used. Also, under present conditions, there are some commercial disadvantages associated with the growing of mixtures, to both the seed trade and the farmer growing grain for malting. Nevertheless, even if satisfactory control is not obtained by mixtures alone, their use in combination with fungicides may prove to be a valuable and durable means of control.

In recent years, some varieties with good resistance to mildew have been released (see *NIAB* Recommended Varieties) but there has been a tendency for the resistance of such varieties to be eroded after a few years of widespread commercial use. The immediate prospect of breeding varieties with satisfactory durable resistance to barley mildew is said to be poor, although the selection of very susceptible varieties can be avoided. In the absence of satisfactory resistance the use of fungicides is often necessary to obtain a good control of the disease.

Fungicides. These were introduced in the early 1970s and have provided a very effective control of the disease. Since the mid-1970s over half of the barley crops in England and Wales have been treated each year. It has been estimated that a single effective treatment in spring barley can prevent about three-quarters of the potential yield loss caused by mildew.

The fungicides can be applied as a seed treatment or a spray. Most of the chemicals are systemic or partially systemic in their action and act to some extent as eradicants as well as protectants.

The choice of fungicide will depend on several factors such as effectiveness (including reference to resistance, see below), cost, the method of application and how this fits in with the management of the crop on the farm, and whether the fungicide is to be used as a prophylactic insurance against mildew or applied only if it becomes necessary to control the disease. Seed treatments are applied as an insurance against mildew and are, therefore, most likely to be used in situations of high risk, e.g. on early-sown winter barley, on spring barley adjacent to winter barley, on very susceptible varieties, on late-drilled spring crops, and in areas or on farms where mildew is consistently an important disease. Seed treatments are also useful as a means of spreading the work load where a large area of barley is grown and may have to be treated for the control of mildew.

Sprays also can be used prophylactically but are mainly used after mildew has appeared in the crop. If the mildew does not appear until late or if it seems unlikely to reach significant levels, then sprays can be withheld and the cost saved.

Spray timing. Sprays check the development of mildew for a period of about 3–5 weeks and sometimes for even longer. The effectiveness of a treatment depends not only on the persistence of the chemical but on the timing of the treatment so that the amount of mildew in the crop (the inoculum) is reduced to, or kept at, a very low level. If this is done, and especially when conditions are favourable for disease development, the mildew takes a long time to build up to damaging proportions and the effect of a single treatment is then very long lasting.

In spring barley it is usually economic to apply one or two sprays so it is important to time them so as to give maximum effect in terms of increased yield. Experimental work has shown that for maximum yield increase from a single spray it should be applied *as soon as* 3 per cent of the leaf area of the third or fourth youngest expanded leaf is affected by mildew. At an early growth stage this means only a few pustules on the small leaves near to the soil and at a later crop stage a very light scatter of pustules on a larger leaf. Some mildew fungicides are compatible with herbicides and it may be convenient to apply both together at the time appropriate for the herbicide provided mildew is present in the crops at that time. More often it is necessary to apply the fungicide later than the herbicide. If there is any doubt about the timing of the spray it should be remembered that delay in applying the fungicide after the recommended stage of disease development is likely to incur higher penalties in terms of yield loss than a slightly earlier than optimum application.

In districts where mildew is consistently a serious problem a programme of two fungicide treatments, a seed treatment and a spray or two sprays, may be economically worth while especially for the more susceptible varieties. Following seed treatment, the spray should be applied as soon as mildew becomes obvious in the crop. When two sprays are used, the first is usually applied with the herbicide or as soon as mildew is first recorded, whichever is earlier; the second spray is applied 3–4 weeks later.

In winter barley, seed treatment and the autumn application of sprays have given good control of mildew but the subsequent effect on yield has been inconsistent, varying from significant yield increases to reduced total yield or a reduced grain size. This latter effect has been associated with a control of mildew in the autumn resulting in excessive tillering especially on heavier soils; here, sprays are not warranted unless mildew is seen on the younger leaves. Sprays in the autumn can be beneficial, even with mildew at low levels, especially on crops growing on light or thin soils.

Sprays applied in the spring to winter barley have tended to give more consistent yield responses though they are less consistent than those in sprays applied to spring barley. Present tentative recommendations are that winter barley which has carried mildew through the winter should be sprayed when mildew is first noticed on the new spring growth. In crops that have escaped autumn and winter infection, and also in crops at later growth stages, the criteria used for spring barley should be applied. That is, apply a spray when about 3 per cent mildew is recorded on the most severely affected leaf.

Winter barley may also be affected in the early spring by other leaf diseases, e.g. the rusts, rhynchosporium, net blotch and by eyespot, so that the control of these diseases as well as mildew should be considered when selecting a fungicide or mixture of fungicides for application to the crop. A broad-spectrum fungicide applied at the stem erect/first node stage often gives good control of the range of diseases including mildew and has been found to give consistently good and cost effective yield responses. The crop should be monitored subsequently because in some seasons a later spray, usually at about flag leaf emergence, may be required to control the late spread of mildew. Again, other diseases may be present and influence the choice of fungicide to be used.

If mildew has not been severe enough by ear emergence for a spray to have been justified, then under most conditions later sprays will not be economically worthwhile. When assessing the cost effectiveness of sprays applied at a relatively late growth stage, the loss in yield due to tractor wheeling damage should be taken into consideration (see page 28).

Fungicide resistance. Some resistance to ethirimol (mainly used as a seed treatment) was recorded in the 1970s though it was not clear whether this seriously impaired the level of control in the field. Observations were confused by failure of control associated with a lack of absorption of ethirimol by the roots under dry soil conditions. More serious was the resistance to the widely used, and formerly effective, 'DMI' fungicides, which was widespread by the mid-1980s. This was associated with a variable and partial loss of control which, in many cases, was commercially unacceptable. At present there are no reports of resistance to the other widely used 'morpholine' group of fungicides. In an attempt to overcome problems of disease control arising from resistance, 'DMI' fungicides are now often recommended or formulated for use in mixtures with fungicides with different modes of action, especially those in the 'morpholine' group (page 27).

In cases where it is necessary to apply more than one treatment to control mildew and other diseases, an attempt should be made to diversify the fungicides in respect of their mode of action, so as to reduce the risk of fungicide resistance developing (see note on fungicides and resistance, page 25).

Leaf stripe (*Pyrenophora graminea* [*Drechslera gramineum* syn. *Helminthosporium gramineum*])

Leaf stripe is now uncommon but before the widespread use of organomercury seed disinfectants it was a damaging disease, especially in the north. The disease symptoms are very characteristic; long stripes develop on the leaves, few in number, often running the entire length of the leaf, at first pale green, then yellow, and finally brown in colour. All or most of the leaves on an affected plant show these symptoms and some may split along the infected stripes giving a shredded appearance to the leaves. Affected plants are most noticeable in the crop at the growth stage immediately before the ears emerge.

The fungus is seed-borne, being present as spores on the surface of the grain or as mycelium in the seed coat. It invades the coleoptile (sheath) as it emerges, passes to the inner part of the sheath and from there infects the first leaf as it pushes through, causing a long green-yellow stripe which later becomes brown. The second leaf is then infected as it passes through the sheath of the first leaf and in a similar manner subsequent leaves become infected. The ear also becomes infected in this way and remains entirely or partially enclosed within the sheath. If it does emerge, some grains remain undeveloped and others are discoloured and infected or contaminated by the fungus. Such ears have a rat-tailed appearance with all the grains thin and shrivelled. Occasionally, as a result of favourable growing conditions, the upper leaves and ears escape infection. The fungus sporulates on the brown stripes. Spores from this source do not infect and cause symptoms in plants in the current crop but serve to spread the fungus to the healthy grain which is the source of the disease in the subsequent crop. The perfect stage (*Pyrenophora*) has not been found in Britain.

Severe attacks may kill seedlings; this phase has not been noticed in recent outbreaks. The disease so reduces the vigour of the plant that even if the ear emerges the yield of grain is negligible. The loss in yield is therefore related to the number of infected plants and can be severe, e.g. 50 per cent stems (ears) affected causes approximately 50 per cent loss in grain yield. Losses as high as 50–70 per cent have been recorded.

Cool weather conditions favour the early development of the disease so that barley sown in the autumn and in early spring is the most severely affected.

Control

The disease is well controlled by some seed treatments. At one time the standard treatment was organomercury and although these chemicals remain generally effective there is some evidence of occasional lack of control associated with the development of organomercury resistant strains. Some of the systemic seed treatments provide an effective control.

Recent isolated outbreaks of the disease were all associated with the use of untreated seed from stocks which had not been treated for two years or more. Occasionally outbreaks occur in crops sown with on-farm-treated seed where the method of treatment does not ensure an adequate cover of fungicide on all the seeds.

The disease could become important again if farmers, saving their own seed, were to abandon regular seed disinfection on a large scale. Untreated seed should not be used unless the crop from which it was obtained had been grown from treated seed or the crop was inspected at ear emergence and found to be free from leaf stripe.

Net blotch (*Pyrenophora teres* [*Drechslera teres* syn. *Helminthosporium teres*]) (Plate 53, page 195)

Net blotch was a common though rarely damaging disease until 1979 when some winter barley crops in the south of England were severely affected. In the following two seasons the disease became increasingly severe and more widespread although the most severely affected crops were concentrated in the south and south-west of England. It is unlikely that the sudden increase in the disease was associated solely with the rapid expansion of the winter barley area which had just taken place because the disease also became more severe at the same time in the well-established areas of winter barley in France and West Germany.

Some of the increase in the disease may be attributed to a series of relatively wet summers. However, probably the most important factor was the introduction at that time of particularly susceptible varieties, and in the United Kingdom of the variety Sonja. Earlier sowing in the autumn and decreased use of the mouldboard plough also favoured the development of the disease from straw-borne sources of infection. Another important factor was the widespread use for a few seasons previously of certain fungicides which gave an excellent control of leaf diseases and eyespot but had no effect on net blotch. The use of such fungicides, especially in the spring, may have encouraged net blotch by making available to the net blotch fungus leaves free from other diseases, thus providing a larger base from which the disease could spread later in the season.

Symptoms

Net blotch symptoms are sometimes confused with those of leaf stripe but they are quite distinct. Net blotch differs from leaf stripe in the shape and size of the lesions and their more or less random distribution on plants and on the leaves as compared with the orderly sequential infection of successive leaves on infected plants in the case of leaf stripe. Furthermore the ear appears normal and is not partially or entirely retained in the leaf sheath as often occurs in leaf stripe.

The net blotch fungus, when seed-borne, infects the sheath (coleoptile) as it emerges and then the first leaf on which it causes a brown stripe extending most of the length of the leaf and usually surrounded by a pale yellow chlorotic area. Spores (conidia) are produced on the stripe and these spread the fungus to other leaves and to adjacent plants. Another symptom associated with seed-borne infection is a greyish lesion in the centre of the blade of the first leaf which later develops into the brown net symptom.

Primary infections of young plants are caused by spores of two kinds (ascospores and conidia) produced on the stubble and straw of a previously infected barley crop as well as by conidia produced on lesions caused by the seed-borne phase. There are three types of leaf symptom, generally described as blotches, spots and stripes. On young plants, the disease usually shows as the characteristic net blotch symptom. Each blotch is irregular in shape, dark brown and on close inspection will be seen to consist of a network of narrow lines, usually surrounded by a yellow chlorotic area: several blotches may run together to form a large lesion. Alternatively the disease may occur as various forms of leaf spot, often with a chlorotic halo, in which the net symptom is not obvious. It has been suggested in Denmark and the Netherlands that leaf spots may be caused by a distinct strain of the pathogen.

Later, and especially after the stem elongation growth stage, a stripe symptom dominates. The stripes are usually short, about 1–5 cm (up to 2 inches) and although some may exhibit the net symptom the stripes often appear as simple brown necrotic lesions, sometimes surrounded by pale chlorotic tissue. In severe attacks the lesions coalesce and can cause the death of the leaf. The disease also spreads to the ear causing small brown necrotic spots but the ear appears normal (in contrast to the symptoms of leaf stripe).

Sources and spread

The seed-borne phase of net blotch is moderately well controlled by organomercury and some other seed treatments but the lack of complete control means that even treated seed can be a source of infection. Although the proportion of infected plants produced from treated seed may be as small as 1 per cent, observations suggest that when circumstances favour spread this may be sufficient to initiate a severe attack.

The major source of net blotch is the straw debris and stubble of a previous infected barley crop. Both kinds of spores, conidia and ascospores, are produced on the straw and are dispersed to infect volunteers and early-sown crops in the autumn. Conidia are produced on the straw over a long period. Perithecia, which contain the ascospores, can be found in the autumn and spring and are more abundant on the cut straw than on the stubble. The main period of ascospore release appears to vary but large numbers have been observed in October. Both conidia and ascospores are dispersed in rain splashes and in air currents. Rapid spread of the disease is through conidia produced on leaves within the crop though there is some evidence of spread from adjacent crops. The fungus does not sporulate immediately or well on the lesions on green leaves but it can produce masses of spores on diseased senescing tissues.

While straw residues into which the crop is drilled are the most important source of disease, straw in adjacent fields can also provide a source of wind-borne infection. Infections established early on autumn-sown crops may lead to moderately severe attacks in the autumn, but usually the disease remains at a low level through the winter and spring though even then there may be a small local flare-up of disease following a spell of mild wet weather. Damaging attacks of net blotch occur mainly during the period after flag leaf emergence, when the disease appears suddenly as numerous lesions on the flag leaf and on one or two leaves below the flag. However, close monitoring of the crop shows that this sudden development is preceded by the appearance of a small number of lesions on the mid to upper leaves. Severe attacks may occur at any time up to crop maturity and in late attacks the ears may also be affected.

Infection is favoured by wet or very humid conditions and lesions appear after 5–15 days.

During wet summers spring barley may also be affected by net blotch, usually when adjacent to a crop of diseased winter barley. In Denmark and New Zealand severe attacks of net blotch in spring barley have been associated with a poor control of the seed-borne phase of the disease when organomercury seed treatments were replaced by less effective substitutes.

Effects on yield

In the epidemics of 1980 and 1981 moderately severe attacks of net blotch in the autumn and early spring did not adversely affect yield. In the same years severe attacks during the summer caused serious yield losses. In a series of fungicide experiments in crops mostly affected with severe disease, control of net blotch gave yield increases in the range 12–40 per cent. Attacks occurring late in the season, after the milky ripe stage, have a negligible effect on yield. The loss in yield is mainly associated with a reduced grain size.

Control

Good control can be obtained by avoiding the more susceptible varieties. Growing successive crops of susceptible varieties greatly increases the risk of severe infection. Net blotch in the autumn is favoured by early sowing and by cultivation methods which do not bury the straw of the previous crop. Early sowing is associated with high yields so that whilst late sowing cannot be recommended, sowing very early in September should be avoided.

Where the disease risk is considered high an attempt should be made to reduce inoculum in the field and adjacent fields by removing or burning the straw and then burying as much of the residue as possible by ploughing. The disease is favoured by high levels of nitrogen fertiliser. Adequate nitrogen is essential for high yields but excessive nitrogen should be avoided.

Some fungicides give effective control of the disease. Spraying affected crops in the autumn has usually not altered the pattern of disease development in the following spring and summer nor has it produced yield benefits. However, if the disease shows signs of spreading to become severe in the autumn (more than 10 per cent of the leaf area affected), then a fungicide spray should be applied as a precaution.

In the spring and summer fungicide treatments should be applied when there is evidence of the disease beginning to spread. The most effective single spray treatment is one applied after flag leaf emergence as soon as a few blotches are noticed on the mid-upper leaves. Additional sprays before or after this spray will improve disease control and yield response especially when attacks are severe.

Where fungicides are applied in the spring (usually at the first node growth stage) for the control of eyespot and a range of leaf diseases, in crops at risk to net blotch, some care should be taken to select a fungicide, or a mixture of fungicides, which are also effective against net blotch—even though the disease may not be noticed at that time. Otherwise, experience indicates that net blotch may be encouraged. The use of an effective fungicide at first node growth stage would not, on its own, control the disease adequately but it may improve the control given by sprays applied after flag leaf emergence.

The standard organomercury seed treatments give good though incomplete control of the seed-borne disease and there is some variation in the effectiveness of the organomercury compounds used.

Some control is also given by the newer systemic fungicides. In New Zealand a seed treatment with triadimenol gave an excellent control at first but then failed, apparently due to the presence of triadimenol-resistant strains probably imported on seed from Europe.

Better control of the seed-borne phase of the disease is desirable but it is unlikely to control the disease in most crops because of the greater importance of straw as a source of inoculum.

Leaf blotch (*Rhynchosporium secalis*) (Plate 54)
This disease has been known for many years but was only noticed as a damaging disease in south-west England in the early 1960s. This was associated with the intensification of barley growing in that part of England and with the introduction of some new varieties which, as it transpired, were very susceptible to the disease. Leaf blotch occurs in all parts of Britain but is consistently troublesome only in south-western areas and in some coastal districts elsewhere. It is common and occasionally damaging on rye and some grasses, especially ryegrasses. The fungus exists in a number of specialised forms which are, in general, restricted in their host range. However, although the form on barley does not infect grasses, under certain conditions the form on grasses may infect barley.

Symptoms
The disease is well described by its North American common name 'scald'. The first signs of the disease are large (1 cm or more), pale green oval lesions on the leaf blade. These enlarge slightly with age, becoming pale brown with a dark brown or purple-brown wavy margin. At this stage lesions can be found in small numbers on the lowest leaves on young plants. Later in the season as the disease spreads, lesions become more numerous and in severe attacks on the upper leaves they may be so closely crowded together that they are small and do not show the characteristic shape and colour. However, the disease can always be identified by the characteristic lesions on the lower leaves. Although the leaf blade is the most severely affected part, the leaf sheath and the ears can also be affected.

Rhynchosporium is often confused, particularly in the early growth stages, with various kinds of non-parasitic leaf spots (see page 199). The distribution of lesions on the plant, especially the fact that some are always present on the older leaves, is a helpful diagnostic feature of the disease. The non-parasitic spots are often confined to the younger leaves.

Effect on yield
In autumn-sown barley the disease is common in the early spring and may spread in the early growth stages. In spring barley the disease is rarely severe in the early stages but in both crops under favourable conditions it tends to build up rapidly after the emergence of the flag leaf. Because severe attacks are relatively late, the disease affects yield almost entirely by reducing grain

Plate 53. Net blotch of barley. Symptoms on young leaves (left) and older leaves (right).
Long Ashton Research Station.

Plate 54. Leaf blotch of barley caused by *Rhynchosporium*. The large blotches have dark wavy margins.
National Institute of Agricultural Botany.

Plate 55. Halo spot of barley caused by *Selenophoma* has many small characteristic angular spots.
Crown copyright.

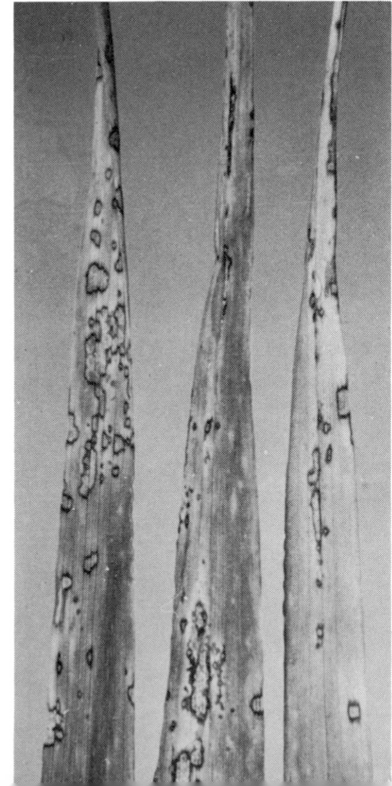

size. Yield losses of about 40 per cent may occur in severely attacked susceptible varieties but losses are usually much lower. The average annual loss in the barley crop in south-west England in the late 1960s was estimated at 2–4 per cent but for England and Wales estimates based on surveys in 1967–80 put the average annual losses at less than 1 per cent. In spring barley a relationship has been established between grain yield and disease levels on the top two leaves at the growth stage when the grain is two-thirds its eventual size, and still soft and wet. The percentage yield loss is approximately equivalent to two-thirds the percentage leaf area affected for the flag leaf, or to one-half the area affected for the leaf below the flag leaf. A good estimate of the percentage yield loss is given by averaging these two figures.

Sources and spread

The fungus can survive for up to one year in debris from the previous season's crop and this is an important source of infection. The disease is common on volunteer plants in the autumn and these are a particularly important source of the disease in southwest England where, because of the mild winter, volunteers often survive in large numbers. Autumn-sown barley crops, especially of the more susceptible 2-row varieties, are also a source of the disease for spring-sown barley. The fungus is seed-borne but the significance of this in initiating epidemics in crops is not known, although it is probably a means of introducing the disease to previously unaffected land. Organomercury seed disinfectants do not eradicate the fungus on the seed. As mentioned above, grasses may be another source of the fungus but their relative importance is not known.

The disease is favoured by cool moist weather. The fungus can produce spores and infect plants at relatively low temperatures but little spread occurs at temperatures above 20°C. The fungus is spread mainly in water droplets, especially during rain, and this accounts for the patchy distribution of the disease sometimes noticed in crops. Although some spread of the disease probably occurs through air-borne spores, most epidemics are initiated from very local sources.

Control

In most parts of the United Kingdom leaf blotch is not of sufficient importance for specific control measures to be advocated. In areas where the disease is important the more severe yield losses may be avoided by the choice of suitable varieties (see NIAB

Recommended Varieties). Some winter and spring varieties now have high levels of specific resistance which is being introduced especially to varieties with the prostrate habit of growth because they tend to be more susceptible to the disease. It is not yet known whether this form of resistance will be durable under British conditions. However under most circumstances avoiding the very susceptible varieties should provide a sufficient means of control.

In spring barley early sowing tends to favour the disease and, although very early sowing should be avoided, late drilling cannot be advocated because of the yield penalties this incurs. Autumn sowing of spring varieties is not advisable since severe infection and severe losses are common under these circumstances.

Autumn cultivations to eradicate volunteers and ploughing aimed at burying crop residues will reduce two important sources of infection but are unlikely to do more than decrease the incidence of primary infections. Subsequent development of the epidemic is then mainly dependent on weather conditions. Although the severity of leaf blotch appears to increase with higher nitrogen applications, it is not advisable to reduce nitrogen fertiliser because the penalties, in terms of yield, are likely to be much greater than the benefits from lower levels of disease.

Some fungicides give good control of leaf blotch. Because the disease is severe infrequently, prophylactic spraying is not economically worthwhile. Satisfactory criteria for the timing of sprays have not been developed but it is suggested that sprays should be applied when the disease is generally distributed but at a low level in crops of very susceptible varieties.

Halo spot (*Selenophoma donacis*) (Plate 55, page 195)
Halo spot of barley is fairly common but is not a damaging disease except on rare occasions in south-west England. The fungus affects grasses, notably timothy and cocksfoot and has been recorded very occasionally on wheat, mainly spring-wheat. Specialised forms exist. The spots are small (1–3 mm) and characteristically square or rectangular in shape, pale brown in the centre with dark brown well-defined margins. On the pale centre of the spots, small black spore-cases (pycnidia) are arranged in lines along the veins of the leaf. The arrangement of the pycnidia, which is a diagnostic feature, may be seen with the naked eye but usually a lens is necessary. Halo spot and leaf blotch (page 194) are often found together but can easily be distinguished by the size of the spots and the presence of the black spore cases on the halo spots.

Halo spot affects the leaf blade and sheath and the ear, especially the awns. The main sources of the fungus are seed, straw debris from the previous barley crop and volunteers. Attacks are rarely noticed in the early stages of the crop but from the end of May the disease may flare up for a short time when characteristically the top leaves are more severely affected than the lower ones. Like leaf blotch the halo spot fungus spreads in water splashes and is favoured by wet weather.

Most of the severe attacks have occurred on newly released very susceptible varieties. Varieties are not usually screened for this disease in breeding programmes. Avoiding such varieties appears to provide sufficient control of the disease though it can also be controlled with fungicides.

Leaf spot (*Septoria nodorum*)

This is an important disease of wheat (page 125) but has also been recorded on barley. Early infections show as distinct brown spots on the leaves but later infections produce irregularly shaped and ill-defined blotches on the leaf blade and sheaths, often at the junction of the two. The black spore cases (pycnidia) of the fungus are often prominent in the centre of the spot. Sometimes a characteristic triangular lesion occurs at the base of the leaf blade just above the ligule. The disease is not known to be important although some varieties are said to be particularly susceptible. It occurs together with other diseases which are favoured by wet or humid weather conditions such as leaf blotch (page 194) and halo spot (page 197). In Scotland the common species on barley is *S. avenae* f. sp. *triticea*. A similar fungus has also been found commonly in England and Wales but mainly associated with leaf tip senescence.

Septoria nodorum is seed-borne and also survives on straw debris. Barley is more susceptible to isolates of *S. nodorum* from barley than to those from wheat and wheat is more susceptible to isolates from wheat, but cross inoculations are successful. The importance of barley straw as a source of infection for wheat crops has not been assessed.

Septoria passerinii has been recorded on barley in the United Kingdom in the past but recently it has not been found and it is possible that it has been previously confused with *S. nodorum*. Another fungus, *Asochyta graminicola*, also produces pycnidia on leaves but this appears to be only a weak parasite of near-senescent leaves and especially leaf tips.

Other barley leaf spots and blotches (various causes)
In the spring, especially in spring barley, pale brown spots can be confused with blotches caused by *Rhynchosporium* but a careful examination of the spots and in particular their distribution on the plant can enable a correct diagnosis to be made (page 194). The pale brown spots at this stage are considered to be non-parasitic in origin and rarely cause serious damage to growth or the yield of a crop. Also in the spring, manganese deficiency can cause numerous pale brown spots on the leaves (page 207).

From about the first node stage onwards an assortment of purple-black to brown spots or blotches can occur, sometimes very numerous and tending to occur most frequently on older leaves, particularly on winter barley plants. The size of the lesions varies from pinpoints, often collectively forming a speckle, to spots and blotches up to 3 or 4 mm long. Dark brown spots can be a resistant reaction to mildew (page 182) and, if so, traces of white mildew mycelium can be seen on them with the aid of a hand lens. Sometimes brown or yellow rust may be associated with brown lesions but again characteristic pustules can usually be detected in at least some lesions with the aid of a hand lens.

Occasionally, a *target spot* symptom occurs—concentric rings of brown tissue separated by pale brown or green tissue. The target sometimes has a marked centre spot which is associated with a mildew or rust infection but frequently there is no pathogen that can be detected or the centre spot is absent. This particular symptom has also occurred under experimental conditions in artificial growth cabinets when it appeared to be associated with a period of high humidity just before ear emergence.

The specific causes of this range of symptoms are not clear but they are often related to variety and the more severe symptoms are frequently associated with plants subject to severe moisture stress, sometimes with high temperatures.

Later in the season the upper leaves of barley plants may show indefinite blotches or speckling associated with heavy deposits of pollen. The lesions are rarely as severe as those which occasionally develop in wheat (page 32). Sometimes anthers which lodge in the axils of the leaves are infected by the fungus *Botrytis cinerea* and this initiates dark brown lesions at the base of the leaf blade upon which the fungus can be seen sporulating very sparsely.

Most of the symptoms described are unlikely to cause serious loss of yield. However, occasionally, when symptoms are severe, they are likely to be damaging though they are often associated with factors (e.g. drought) which are themselves responsible for yield losses.

Barley yellow dwarf (Barley yellow dwarf virus – BYDV)
This virus disease affects cereals and grasses and is spread by aphids. The disease causes obvious, sometimes severe, symptoms in cereals though wheat is less sensitive than barley and oats. In grasses the symptoms are rarely noticed by casual observation. BYDV was first recognised in the United Kingdom in 1954 and although significant outbreaks occurred in 1958 and 1971, serious outbreaks were not prevalent until the adoption of early (during September) drilling practices in the 1980s.

Symptoms
Barley displays a wide range of symptoms depending mainly on the time of infection but also on the strain of the virus. Typical symptoms are a bright yellow colouring of the leaf extending from the tip towards the base. Leaf yellowing is a common response in barley to various disorders, but BYDV-infected plants may be distinguished from others by the particularly bright yellow leaf colour and the fact that infected plants may be growing amongst, or adjacent to, normal green plants. With the exception of some autumn infections, all the leaves produced after infection have the yellow symptom. Thus on some plants the lower leaves are green and the upper leaves show the yellow symptom. With late infections only the youngest leaves, often only the flag leaf, may show symptoms. In other barley disorders which cause yellowing, the lower leaves tend to show symptoms first and the flag leaf last.

Plants affected in early autumn with severe or mild strains of the virus may be killed during the winter or, if they survive, remain very stunted and produce few or no fertile tillers. Plants infected later in autumn and winter produce fertile tillers but these tend to be stunted, being 20–40 per cent shorter than healthy plants. Plants affected at this time produce bright yellow leaf symptoms in late April-early May just before the period of stem elongation and it is at this stage that infections in winter barley are most obvious. Leaves produced subsequently may not show the typical bright yellow symptoms. Instead they are green but often display alternating pale and dark green stripes along the length of the leaf. However, some plants infected in the autumn-winter period, possibly the ones infected later, will show the more typical leaf yellowing symptoms.

Plants infected in early autumn may show a yellow leaf

discolouration during late October and November but more usually the symptoms are not seen until the spring when affected plants occur in distinct patches. When attacks are severe the patches merge to give large areas of affected plants.

In some winter barley varieties the awns of affected plants are distinctly more purple than those of healthy plants.

Plants infected in spring or early summer are usually scattered at random through the crop and show the typical yellow leaf symptoms. Affected plants may be slightly stunted but often they are not noticeably different in size from healthy plants.

Strains of BYDV and aphid vectors

The virus exists in a number of strains differentiated by the severity of the symptoms they cause in cereals. Also there is a relationship between the strains and the aphid vectors (carriers) so that some strains can be transmitted efficiently only by specific aphids.

BYDV is introduced into cereal crops and spread within crops by aphids. The two most important vectors of the virus in the United Kingdom are the bird-cherry aphid (*Rhopalosiphum padi*) and the grain aphid (*Sitobion avenae*) (see also pages 60–62). The rose-grain aphid (*Metopolophium dirhodum*) is considered to be much less important but it can occur in large numbers in the summer, when it may cause direct damage, and then may be of some importance for within crop spread.

The strains of BYDV are sometimes referred to specific types, first described in the USA, such as RPV (most severe), PAV and MAV (least severe).

In the United Kingdom the most severe strains are of the RPV type and are spread only by the bird-cherry aphid. These strains appear to be relatively uncommon. Rather less severe strains of the PAV type are common and are spread by both bird-cherry and grain aphids though the former are the more efficient vector. The 'mild' strain (MAV type) is transmitted almost exclusively by the grain aphid. The rose-grain aphid transmits mild strains, some of which may be different from the MAV type.

The terms 'severe' and 'mild' are entirely relative and under some conditions the so-called 'mild' strain (MAV type) can be very damaging. For example in the mid-1980s in Yorkshire this strain, transmitted by the grain aphid, caused symptoms in winter barley ranging from the death of plants to slight stunting, depending upon the time of infection. These cases illustrated clearly that the severity of symptoms is not only an expression of the strain but also of the time of infection (with earlier infections causing most damage).

The distribution and prevalence of strains of BYDV varies from one season to another depending on the numbers and activity of the aphid vectors. Usually the strains from the western and southern parts of the United Kingdom are of the more severe type and are mainly associated with the bird-cherry aphid. Strains from the north-eastern parts are of the mild type and mainly associated with transmission by the grain aphid.

Sources of the virus and spread

The main sources of BYDV are grasses, wild and cultivated, which are commonly infected though often without symptoms, and other cereal crops or cereal volunteers. There are two ways in which the virus can be introduced into a cereal crop:

1. By non-migrant wingless aphids which are already present in the field on grass or on cereal volunteers and then move on to newly emerging crops. The 'green bridge' transfer of the virus occurs when insufficient time elapses between ploughing or other cultivations and sowing for the elimination of aphids on the surviving grass and cereals. This form of the disease is most common in the south-west and western parts of the country and is mainly associated with the bird-cherry aphid. BYDV from this source is potentially very damaging because infections are established in the young, newly emerged seedlings. Severe attacks are seen mainly in autumn-sown crops but have also been recorded in spring barley.

2. By winged aphids migrating into crops from grasses or cereals elsewhere. This is the most common source of the virus.

In the autumn migrating aphids alight in cereal crops until the end of the migration period in late October. The earlier the crop is drilled, the more likely it is to be infested by migrant aphids; crops sown in August and early September are particularly affected. Crops sown after mid-October normally emerge too late to be infested to a significant extent and are therefore little affected by BYDV.

When the virus is introduced into a crop in late spring or summer, infected plants tend to be isolated and occur at random in the crop, though there may be a few small patches. Spread during the summer may be too late for symptoms to appear, but there is evidence that the incidence of such infections in some years can be quite high.

Once established in the crop the aphids reproduce and the offspring spread to neighbouring plants. If the aphids carry the virus then this is spread at the same time to produce the characteristic patches of infected plants. In very early drilled crops significant spread may occur during September and October but in most cases spread occurs later. The amount of spread during the winter is dependent on the weather and its effect on aphid activity. Severe weather which kills most aphids will stop spread. In the absence of such weather, spread occurs periodically during the late autumn and winter months when breaks in the weather favour aphid activity.

Effect on yield

The effect of BYDV on cereal yields is very variable since it depends on many factors, the most important of which are the strain of the virus, the time of infection and the rate of spread. Infection of young plants causes severe stunting and sometimes death. Later infections cause progressively less stunting and some plants may be of normal size but show the characteristic leaf symptoms.

Early infections of cereals from sources within the crop, usually ploughed-in grass, can lead to very serious losses, occasionally the loss of an entire crop. More usually the severe effects of early infections which lead to complete loss of grain are limited to small areas. Following infection of winter cereals in the autumn–winter, grain losses in affected areas are usually up to 50 per cent while infections in the early spring cause correspondingly less damage, up to 30 per cent. The overall effect on the crop depends on the number of affected plants as well as on the time of infection. Although large areas are sometimes affected, the disease usually remains in discrete patches which often occupy not more than 10 per cent of the field area and usually much less.

Infections which occur from late spring are not seen until July and very late infections may escape notice because symptoms are obscured by leaf senescence as the crop matures. Such late infections have a very small effect on yield.

Although no surveys of yield losses in the national crop have been undertaken, it seems unlikely that they have exceeded 1 per cent even in seasons when the disease was above average. The position may have changed during the 1980s when a much higher proportion of the autumn sown crop has been sown before mid-October. However, an assessment of the damage caused by BYDV under these conditions has not been possible because of a widespread use of insecticides in the autumn to control the aphids and consequently the spread of the virus.

Control

Crop hygiene and good cultivation practices are normally sufficient to eliminate sources of BYDV within a field. However in some areas, such as the south-west, where ploughed-in grass is sometimes an important source of infection, it may be necessary to take special precautions to ensure that aphids do not survive until the new crop emerges. One recent recommendation is that a desiccant herbicide should be applied 7–10 days before cultivating and leaving 2 weeks between treatment and sowing. As an alternative the application of an insecticide at crop emergence will reduce infection but is not as effective as the herbicide treatment.

The introduction of BYDV into autumn sown crops by migrant aphids can be largely avoided by sowing after mid-October so that the crop emerges after the migration has ceased. However, such practices may not be acceptable because they are likely to result in lower yields. For crops sown earlier, nothing can be done to prevent aphids alighting in the crop and introducing the virus. Control measures are therefore aimed at killing the aphids before a significant amount of spread occurs.

In areas where BYDV is known to be damaging in at least some years, it may be necessary to apply insecticides as a routine measure to crops drilled before mid-October. In other areas it may be possible to restrict spray applications to crops where there is a high risk of infection. Attempts to define 'high risk' are still being developed. They include the monitoring of aphid activity in crops and the use of an infectivity index. An infectivity index is based on the numbers of aphid vectors caught in suction traps and the proportion of these found to be carrying the virus. So far the method has proved useful in assessing the risk of the bird-cherry aphid introducing virus into autumn sown crops. Methods for assessing the risk of infections associated with the grain aphid are still under investigation.

The timing of the aphicide spray is important. It can be limited to a single application provided this is made after the flights of aphids have ceased and before significant spread in the crop has occurred. Experiments have shown that the optimum timing under most conditions is the last week in October or the first week in

November. Occasionally there may be need for an earlier spray in very high-risk situations, e.g. an extremely early infection with spread commencing September, when a spray may be necessary in mid-October.

The treatment of spring cereals to control BYDV is rarely warranted except, possibly, where the virus is common in autumn cereals and there is evidence of over-wintering aphids multiplying in the spring and early summer. Under these conditions late sown crops are particularly at risk.

All cereal varieties are susceptible to BYDV but some show some degree of tolerance, i.e. although they are as readily infected by the virus, their growth and yield are less seriously affected. Sources of higher levels of tolerance and also of resistance are known and are now being incorporated into commercial varieties.

Barley yellow mosaic (Barley yellow mosaic virus—BaYMV)
Barley yellow mosaic virus is so far confined to winter barley. The first sign of the disease is the appearance of patches of yellow or pale green spiky-looking plants usually in February or March. The disease is spread by a soil-borne fungus, *Polymyxa graminis* and can cause yield losses of up to 40 per cent. It was first recognised in Japan in 1940 but was not recorded in Europe until 1978, in West Germany. Subsequently it has been found in England and France.

The first confirmed case in England was at Abingdon (Oxon) in January 1980 and this was soon followed by others so that by the end of 1981 over 200 cases had been reported and there was a further increase in subsequent years. Most of the outbreaks were in East Anglia but almost no arable area was free from the disease. Some farmers reported that they had noticed similar patches for up to five years before BaYMV was confirmed so that it is likely that the virus had been introduced some time before 1980.

The symptoms vary during the season and with variety. In February and March affected plants have chlorotic streaks (1–5 mm in length) running along the length of the leaf giving it a mottled appearance. All leaves on an affected plant show symptoms but these are most distinct on the youngest ones. The leaves of affected plants are inrolled and remain erect giving the plant a spiky appearance. Later the chlorotic streaks become necrotic and brown, and in some varieties there is an orange-yellow discolouration followed by death of the oldest leaves. When symptoms are first seen there is no obvious difference in the height of affected and healthy plants but as plants grow and stems elongate in April, affected plants grow more slowly and produce fewer fertile tillers. In April and May the chlorotic streaks in the younger leaves become less distinct and may have disappeared by June, though often the effect on plant size persists. The expression of these symptoms in early spring and summer is related to temperature. Cool weather favours more severe symptoms which may disappear after a warm spell only to return if temperatures drop again. The severity of symptoms appears to be related to the amount of virus in the plant and tends also to increase if plants are under stress, for instance in adverse soil conditions.

BaYMV infects only barley and closely-related wild species but symptoms have so far been seen only in autumn-sown barley crops probably because of the temperature relationships.

The virus is introduced into crop plants when their roots are infected from spores of the soil-borne fungus *Polymyxa graminis* that are carrying the virus. This fungus is not a damaging pathogen except as a virus carrier. It is common in cereal roots and fortunately few populations of the fungus appear to carry the virus. Resting spores of *P. graminis* are confined to the soil in which they can survive for many years and it is known that the virus can also survive in the absence of barley for five years and probably for much longer. The resting spores of the fungus germinate and give rise to single-celled zoospores which swim in soil water and infect roots of the newly planted crop. Spread from field to field may occur by movement of soil or crop debris on farm implements etc., but how important this might be is not yet clear.

The extent and distribution of infected patches varies. In some outbreaks whole fields of up to 20 ha were uniformly affected and in others the disease occurred in variable numbers of various sized patches. In some cases these appeared to be randomly distributed but in others the disease was associated with ancient field boundaries, with the line of a grubbed hedge, with pathways through fields or with gateways. Some patches are kite-shaped possibly suggesting spread in the direction of primary cultivations. Aerial photographs taken in two successive years have indicated that the patches may increase in size two to five times. There is some evidence that spread occurs more rapidly where tine cultivations have been used, presumably as a result of dragging infected plant material across a field.

Control
Some varieties of winter barley have good resistance to BaYMV and sources of resistance are being incorporated into new varieties in breeding programmes. In trials a varietal reaction has sometimes varied depending on the preceding crop. Thus some varieties which appear to be resistant can be susceptible when grown as a second successive crop. This may suggest that specific strains can be selected by some resistant varieties.

Where the disease has been confirmed, resistant varieties should be grown or winter barley should be avoided. In the absence of suitable resistant varieties and if it is necessary to crop the land with barley then spring varieties should be grown.

The virus and fungal vector survive on root debris. Cultivations designed to minimise the movement of crop debris will help to restrict the spread of the disease and removing soil from machinery, for instance, by high-pressure water jet, may restrict field-to-field spread. There is some evidence that such methods have successfully restricted on-farm spread.

Manganese deficiency
This occurs most commonly on peaty soils or mineral soils with a high organic matter or a high lime content. Symptoms are seen in young plants at the late tiller stage and later when patches of plants become pale yellow-green and often assume a floppy appearance. Many small pale brown spots develop which may coalesce to form long lines on the leaves. Large areas of the crop may be affected and, although symptoms may later disappear, the deficiency is likely to cause yield reductions. A description of the disease in oats and control measures are given on page 238.

Covered smut (*Ustilago hordei*)
This disease is the equivalent in barley of bunt in wheat (see page 136) and is just as infrequent in Britain. It is, however, much more easily seen in the field than in wheat bunt, because the spore-balls are only held by the thin transparent skin of the grain and the black spore mass is often exposed. The infected ears remain reasonably intact, however, and are easily recognised by their grey-black colour and apparent lack of awns which shrivel soon after flowering. All the ears on a plant are affected as a general rule, as are all the grains in the ear. The spores are not easily wind-dispersed even when exposed, because they are made rather sticky by a coating of an oily substance which also waterproofs them, so that the majority of spores are retained in the ear

until harvest. The combine harvester threshes the infected ears and the spores are distributed over the healthy grain, closely following the behaviour of those of the wheat bunt fungus. Since the spores are held on the outside of the seed grain as a surface contamination, covered smut is controlled by seed treatment with organo-mercury and other fungicides in the same way as bunt (page 136).

Loose smut (*Ustilago nuda*)
This disease is very similar to loose smut in wheat. A detailed description of the disease is on page 138. The barley form of the fungus is distinct from the one on wheat, and itself consists of a number of different races.

The infected plant is almost indistinguishable from healthy plants until near ear emergence, when the careful observer may notice individual flag leaves showing a marbled brownish appearance. Occasionally the leaf sheath may be seen to show a large brown papery blotch at the position of the infected ear in the boot stage. More frequently the first sign of disease is seen as the ear emerges to show the typical sooty spore masses replacing the grain.

The level of attack varies from season to season and, although the trend over a sequence of seasons is for the disease to increase in a seed stock, incidence in any year fluctuates according to the weather conditions during the flowering period of the crop from which the seed was produced. Flowering is protracted in cool moist conditions, allowing greater opportunity for the loose smut spores to infect the ovary. Seed produced in cool, wet seasons is, therefore, liable to carry a higher level of infection than that produced in seasons which are warm and dry.

Variety also influences the degree of attack. Varieties can be classified according to their flowering habit which determines their liability to infection. The open-flowering habit, in which stamens and pistil are exposed, leads to greater susceptibility to loose smut. The 'closed' flowering habit provides a mechanical barrier to the entry of loose smut spores and varieties exhibiting this character are relatively much less susceptible.

The warm-water treatment used for wheat is equally satisfactory for barley except that the temperature is 52°C for fifteen minutes. This treatment is expensive and inconvenient and has been replaced by the use of systemic fungicide seed treatments. The control of loose smut in some stocks of winter barley by one of the systemic fungicides, carboxin, has not been satisfactory in recent years and this may be associated with the development of carboxin-resistant strains. At present there are no reports of resistance to the 'DMI' fungicides used as seed treatments (page 26).

The farmer will normally rely on seed produced under the Ministry of Agriculture, Fisheries and Food UK Seed Certifici-ation Scheme (see page 140) which stipulates a permissible (low) level of loose smut infection. Such seed is from crops normally grown from seed treated with a fungicide effective against loose smut. There is, therefore, no need for a farmer to treat the seed for the control of loose smut unless he intends growing on the stock for seed for himself.

There may be up to a twentyfold increase in loose smut in a seed stock in one season. The actual increase will depend largely on the weather at flowering time but this potential increase should be borne in mind by farmers saving their own seed. Farmers should always inspect and make counts in their seed crops promptly at flowering time; delay makes counting more difficult because spores are quickly dispersed and infected ears are more difficult to detect.

The level of infection in a seed stock can be determined by laboratory examination of a sample of the seed. This involves a staining technique coupled with the dissection of the embryo and gives a reliable indication of the level of infection to be expected when the seed is grown on, field infection usually being about two-thirds of that shown in the laboratory test. The test is done, for a fee, at the Official Seed Testing Stations (page 30). The test is not satisfactory for wheat.

Ergot (*Claviceps purpurea*) (Plates 44, 45, page 141)
Ergots are hard black fungal bodies which replace grains in the ear. They contain poisonous substances and contaminated grain should not be used for feeding livestock. Sometimes grain may be contaminated with ergots from grass weeds in the crop. Barley is less susceptible than some other cereals (a fuller account is given on page 140).

Botrytis disease of spikelets (*Botrytis cinerea*)
Infected spikelets in wheat and barley are pale brown and are particularly conspicuous when the ears are still green. Occasional ears at random in the crop are affected and rarely more than 1–3 spikelets per ear, also at random, show symptoms. The symptom may be confused with those caused by frost (page 210) and high

temperatures in barley (page 211) but can be distinguished by the distribution of affected ears and spikelets.

The fungus infects the spikelet through the dying anthers soon after flowering and the young developing grain is destroyed leaving empty 'glumes' (actually lemna and palea and sometimes the glumes in wheat) which have an oval pale brown lesion occupying most of their surface area. Anthers are frequently colonised soon after flowering by black moulds including *Cladosporium herbarum* which is common on dead cereal tissues (page 145). This fungus is antagonistic to *Botyrtis cinerea* and it would appear that the disease occurs on the infrequent occasions when *Botyrytis* becomes established in the anther before *Cladosporium*. The grey mould of *Botrytis* can often be seen on affected tissues. Although *Botrytis* has also been associated with a rot following damage by thrips, this is thought to be unusual. Sometimes a similar disease of individual spikelets is associated with infection by *Fusarium* spp. though this is distinct from ear blight (page 97).

Black mould

This is a discolouration of the ears by black 'sooty' moulds (see under wheat, page 145).

Black point

A dark shrivelled area on the grain just at the embryo end (see under wheat, page 152) makes the grain unacceptable for malting or seed.

Frost damage

Soon after the seedlings have emerged, parts of the young leaves towards the tips or occasionally whole leaves may turn white, being devoid of clorophyll. Such symptoms often occur in large patches in lower-lying parts of the field and are probably due to direct frost damage, although cold winds are said to cause similar symptoms. Affected plants soon recover. The symptom can be confused with those of boron toxicity ('white strike') which occurs when a boronated fertiliser, as used for sugar beet and swedes, is mistakenly applied to cereals.

Later frosts may cause death of florets resulting in blindness, when some of the spikelets are bleached and do not produce a grain. Flowers do not mature all at the same time so that at any given time only some of them are sensitive to frost damage and this results in only a few spikelets in a particular part of the ear being affected. Sometimes the damage is confined to the tip of the ear, possibly because only the flowers in this part were exposed to frost as the ear emerged from the sheath.

Other conditions, including pest or disease attack or adverse soil conditions, may result in blindness, but in these cases the damage is less well defined and more often than not the whole plant also tends to show symptoms of the condition. There are, however, some pests which produce damage very difficult to differentiate from frost injury (see page 156).

Blindness

Blindness can be caused by frost, copper deficiency or other factors but occasionally some ears show a few blind spikelets, especially noticeable when the ear is still green, with a transparent appearance. This appears to be due to lack of pollination and fertilisation. The spikelets remain open with the gaping bracts and sometimes pollination occurs later and a normal grain develops. When pollination fails completely and the grain does not develop the spikelet is thin and shrivelled amongst the normal developing grains. This symptom has been associated with abnormally high temperatures (and also with low temperatures) during a critical stage in flower development but in some cases, especially when the symptom is common, there is no satisfactory explanation (see also *Botrytis* disease page 209).

Copper deficiency

Symptoms are usually seen after ear emergence. Awns of affected plants are white and may be shed. In the case of severe deficiency the ears are partly or completely blind and often remain erect when the ears of normal plants have bent over. Straw may be weakened and bend or break beneath the ear. The 'blackening' symptom described for wheat does not occur in barley which is less affected than wheat. Symptoms may not be obvious and in some experiments where no symptoms were recognised the application of copper to the crop increased yields by up to 0·8 tonnes per hectare. A more detailed description of copper deficiency and control measures have been given for wheat (page 149).

Herbicide damage (Plates 48a,b,c page 151)

Growth regulator herbicides applied earlier or later than recommended can damage barley in a variety of ways.

Spraying earlier than recommended can cause malformation of ears and, less obviously, in some of the earlier leaves the formation of tubular ('onion') leaves. Malformation of the ears may take several forms: spikelets may be opposite instead of alternate, more numerous, blind or bunched; the rachis (stem of the ear) may be extended in parts so that spikelets appear to be missing, or twisted with some spikelets at right angles to others. When the rachis is extended it is also weakened and may break. Sometimes the emerging ear is trapped in the flag leaf which is fused into a tube and this results in a buckling of the stem below the ear. Mild ear deformity does not result in yield reductions but severe malformation can do so.

Spraying later than recommended does not result in ear malformation but can cause appreciable reductions in yield with poorly filled ears which at ripening are often affected by the black moulds (page 145). The most susceptible stage is at flowering and herbicides applied at this time can prevent proper grain development. These symptoms, which are seen more often in winter wheat (page 152), are similar to those caused by take-all and in fact plants damaged by herbicides are predisposed to attack by the take-all fungus.

Chapter 6

OAT PESTS

PLANT DAMAGE SYMPTOMS

a. *Seed or seedling damage at or before emergence*

Symptom	Cause
Seeds missing from drill row	Birds (p. 75) Rats, mice (p. 77)
or hollowed out	Slugs (p. 68) Wireworms (p. 219) Mice (p. 77)
Seed husks left on soil surface	Birds (p. 75)

b. *Shoot develops but does not reach soil surface*

Symptom	Cause
Shoot short and swollen, root tips clubshaped	Seed treatment injury (p. 23)
Shoot brown and shows signs of feeding	Slugs (p. 68) Leatherjackets (p. 219) Wireworms (p. 219) Frit fly larvae (p. 216)
Shoot long and often twisted	Deep sowing (p. 84) Soil capping (p. 86)

c. *Seedling or tillering plant damaged*

i. Whole plants affected, usually yellow first

Symptom	Cause
Plants pulled out and left on soil	Birds (p. 75) Rats, mice (p. 77)
Plants bitten near soil level	Slugs (p. 68) Leatherjackets (p. 219) Wireworms (p. 219) Grass moth caterpillars (p. 49) Swift moth caterpillars (p. 49)

Symptom	Cause
	Cutworms (p. 49) Chafer grubs (p. 57) Wheat shoot beetle larvae (p. 66) Millepedes (p. 66)
Plants not bitten	Cereal cyst nematode (p. 221) Frost damage (p. 86) Deep sowing (p. 84)
Shoot swollen or mis-shapen	Stem nematode (p. 226)
Plants slightly stunted, leaves spotted or flecked	Aphids (p. 220) Thrips (p. 65)
ii. Centre shoot yellows and dies (deadheart) Outer leaves stay green	Frit fly larvae (p. 216) Grass moth caterpillars (p. 49) Common rustic moth caterpillars (p. 54) Grass and cereal fly larvae (p. 43) Bean seed fly larvae (p. 42) Wireworms (p. 219)
iii. As above, but a small neat hole often present near shoot base	Flea beetles (p. 164) Wheat shoot beetle larvae (p. 66)
iv. Leaves bitten, often well above ground Long narrow holes	Slugs (p. 68)
Roughly circular holes	Grass moth caterpillars (p. 49)
v. Leaves neatly cut off Horizontal bitten edge, tips missing	Mammals (p. 77)
V-shaped leaf bite, tips often lying on soil	Birds (p. 75)
Leaves torn with ragged ends	Slugs (p. 68) Leatherjackets (p. 219)

d. *Plant damaged after tillering stage*

Symptom	Pest
i. Whole plant sickly, yellow or reddish Root system short, much branched, white cysts visible June onwards.......	Cereal cyst nematode (p. 221)
Root system normal..........................	Oat spiral mite (p. 220) Grass and cereal mite (p. 66)
ii. Some shoots yellow, others healthy	Grass moth caterpillars (p. 49) Common rustic moth caterpillars (p. 54)
iii. Some shoots swollen, others healthy	Stem nematode (p. 226)
iv. Stem surface near upper nodes pitted with saddle-shaped depressions	Saddle gall midge (p. 219)
v. Long narrow strips eaten from leaf blade	Slugs (p. 68) Cereal leaf beetle (p. 219) Barley flea beetle (p. 164)
Irregular areas bitten from leaf edge	Grass moth caterpillars (p. 49)
Discoloured blotches on leaf surface	Aphids (p. 220)
Silvery marks on leaf surface	Thrips (p. 65)
Leaves with blister mines..................	Cereal leaf miner larvae (p. 161)
vi. Shoots bend or break and fall over Shoot cut off with diagonal bite or peck mark, heads often stripped.............	Birds (p. 75) Mammals (p. 77)
Shoot breaks off near upper nodes	Saddle gall midge (p. 219)

e. *Damage to flowering heads*

Symptom	Pest
Distal florets white, distorted or shrivelled .	Frit fly larvae (p. 216)
Whole panicle white or aborted	Grass and cereal mite (p. 66)

Panicles sticky, often covered with extruded fluid..	Aphids (p. 220)
f. *Damage to ripening grains*	
Grains missing......................................	Birds (p. 75)
Grains shrivelled or incompletely developed	Grain aphid (p. 220) Thrips (p. 65) Birds (p. 75)
Grain blackened and/or partly eaten	Rustic shoulder knot moth caterpillar (p. 54) Frit fly larvae (p. 216)
Grain replaced by black powdery mass	Frit fly larvae (p. 216)
Grain flattened, white deposit on outside ...	Birds (p. 75)

The range of pests attacking winter and spring oats is rather different from that which affects wheat and barley. Thus oats are immune to wheat bulb fly, late wheat shoot fly, gout fly and Hessian fly. Conversely, stem nematode is an important pest of oats and rye but does not affect wheat or barley. Those pests which attack all four cereals (e.g. cereal cyst nematode, cereal aphids) often produce most violent symptoms and greatest loss of yield in oats.

INSECTS

Bean seed flies (see page 42)

Grass and cereal flies (see page 43)

Frit fly (*Oscinella frit*)

The frit fly has been aptly described as 'a denizen of grassland and a pest of oats'. It is best known for its damage to spring oats, but in addition it can damage maize and autumn-sown cereals as well as many grasses of commercial importance.

Damage symptoms

Shoot damage is seen in spring oats in late May and June and in autumn-sown oats from November until February. Plants up to the four leaf stage are vulnerable. Very young plants may be killed outright, whereas in older plants the central leaf yellows and dies ('deadheart'). The outer leaves remain green and encourage new tillers to be produced. These may in turn be attacked, and further tillers are formed so that eventually the oat plant has a prostrate, grassy appearance. Within each attacked shoot are one or more maggots, each a white, thin, legless grub about 0·6 cm long when fully grown. At maturity, the larval skin hardens and turns reddish brown to form the puparium, which is the stage commonly seen when attacked shoots are dissected.

The differences between frit fly damage and that caused by stem nematode will be noted later (p. 226).

Oat plants are again attacked after tillering has ceased and the stems are beginning to lengthen. Larvae may then destroy the whole or part of the inflorescence while it is still within the stem. The florets chiefly towards the top of the panicle are twisted, white or withered. As the ear emerges, the brown puparial cases can often be found amongst the affected florets.

Panicle damage to oats may be caused by other agencies. The whole ear may be affected by oat spiral mite (p. 220) or incorrect timing of herbicide application (p. 240); the basal spikelets only may be blinded by oat spiral mite (p. 220) or by 'blast' due to a physiological upset (p. 240), or the top spikelets may become blind as a result of late frost injury (p. 240). It is necessary to search for puparia in the spikelets to distinguish frit fly damage from frost tipping.

Grubs of this 'panicle' generation of frit fly also feed on the developing grains but leave the husks intact. The kernels are either poorly filled or completely blackened and aborted ('fritted oats'). Careful removal of the husks will usually reveal the brown puparial cases to confirm that frit fly was responsible.

Life history

The number of generations completed each year varies according to latitude. In southern England there are three or four generations, whereas in northern counties and in Scotland there are no more than two.

Under typical conditions in southern England, adult flies emerge from infested grass and cereal shoots from early May until June, with a peak in late May. Both sexes are small active flies, black and shiny in appearance. Eggs are laid on or near the shoots of grass or oat seedlings, a favoured site being under the edge of the leaf sheath. Maggots quickly hatch from these eggs and bore into the centre of shoots to feed on the tissues there for about three weeks. The full-grown maggot changes into a brown puparium, from which adult flies of the next generation emerge.

These flies are on the wing in mid-July (later in the north of the British Isles) when late-sown spring oats are flowering. Eggs are laid either on young oat plants (e.g. late oats grown for silage) or on the spikelets, the larvae feeding on the developing kernels. Pupation takes place within the oat panicle, and adult flies emerging in late summer often swarm in a halo around oat stooks or congregate on granary windows. The female flies lay their eggs on grasses or volunteer cereal plants or perhaps directly on early-sown winter cereals. The maggots continue feeding within the cereal shoots throughout the winter period, and do not pupate until late February or March.

If grass leys or grassy cereal stubbles are infested with frit fly larvae at the time of ploughing in early autumn, the grubs often leave the rotting turf and migrate upwards to attack a newly sown crop of grass or winter cereal. An interval of at least four weeks between ploughing and sowing will ensure that most of the grubs die in the rotting turf.

Control

Early sowing is the best means of avoiding damage in spring oats, with the end of March the deadline for safety. For every day's delay in sowing after 16 February in south-west England, there is a yield penalty from frit fly damage of 0·3 per cent. The peak danger period for sowing oats in that area is 11–24 April. A well-prepared seedbed, adequately manured, will help the plant to tiller rapidly and after passing the four-leaf stage it is out of danger.

Insecticidal sprays have given some protection to late-sown spring oats against first-generation adults and newly hatched larvae. DDT was formerly recommended for this purpose, but is no longer permitted. Of the less persistent alternatives subsequently tested, chlorpyrifos, omethoate and triazophos have given useful results. Seed treatments would be acceptable for convenience and on financial grounds but so far no insecticides have given reliable results when formulated this way.

Early sowing goes some way towards minimising damage to the grain by the later 'panicle' generation larvae. Here, sprays of insecticides have not proved cost-effective on an experimental plot basis.

Although spring oat cultivars differ in their susceptibility to shoot and probably grain damage, such differences are not sufficiently great to be exploited by plant breeders.

As with other 'ley pests', frit fly damage to autumn-sown cereals is best avoided by ploughing grass or grassy stubbles in late August and not drilling for at least four weeks. Cereal stubbles infested with grass weeds should be cultivated and/or treated with a suitable herbicide (e.g. low-dose paraquat) as soon as possible after the crop has been harvested. The risk of damage to autumn-sown crops does not justify routine prophylactic chemical treatment. However, crops at risk should be examined soon after brairding, and a spray application of an approved insecticide made as soon as damage is first seen.

Leaf miners (see page 45)
Attacks on oats are much less frequent than on wheat and barley.

Saddle gall midge (see page 47)
Eggs are laid on oats but wheat and barley are more favoured hosts.

Leatherjackets (see page 162)
Young oat seedlings are very susceptible to leatherjacket damage.

Moth caterpillars (see page 49)
The range of caterpillars affecting wheat and barley is also found in oat crops.

Wireworms (see page 55)
Oats are more tolerant of wireworm damage than wheat or barley. Oats are occasionally damaged by the larvae of chafer beetles (page 57), wheat shoot beetle (page 58), wheat flea beetle (page 59), or barley flea beetle (page 164).

Cereal leaf beetle (*Oulema melanopa*)
Attacks are usually worse on oats than on other cereals in the British Isles, but the pest is not as damaging here as it is in the USA.

Damage symptoms
Both the adult beetle and its larva eat long thin strips of tissue from the upper surface of the oat leaf, so that when damage is extensive the crop has a whitish appearance.

Life history
The adult beetle is a small insect with a black head, red thorax and metallic blue-green wing cases. It shelters in hedge bottoms

and similar habitats during the winter, then moves into cereal fields from April onwards and often feeds in groups with other adults. The female beetles lay eggs in June or July and the young larvae hatch in about ten days. The grub is sedentary, brownish in colour and hump-shaped. Its slug-like appearance is accentuated by the slimy covering over the body. On close examination the larva is seen to bear three pairs of tiny legs.

After feeding on the cereal leaves for two to three weeks the larvae enter the soil to pupate within a cell a few centimetres below ground. Within three weeks adult beetles emerge to feed until autumn on grasses and cereals. They then migrate to the over-wintering sites.

Control

Attacks are rarely serious enough in Britain to warrant chemical control measures

Natural control is exerted by at least one insect parasite, which is present in European countries but not in the mid-western states of the USA or Ontario. This may explain why cereal leaf beetle is such an important pest in the New World but not in Britain or continental Europe.

Aphids and leafhoppers

The species of aphids which affect wheat and barley (see page 60) also colonise oats. Symptoms of direct injury and of barley yellow dwarf virus infection are often most severe on oats.

Thrips (see page 65)

MITES

Oat spiral mite

Steneotarsonemus spirifex attacks oats and grasses more frequently than wheat in this country. The symptoms are similar to those already described for wheat (page 66) but attacked oat plants are often reddish in colour and the leaves have rolled edges. Damage is more prevalent in hot dry summers, in late-sown crops, and when oats follow oats in the rotation. The damage to the oat panicle can be mistaken for frit fly or herbicide damage (but see page 217). Chemical control methods have not been tested in the United Kingdom.

MILLEPEDES (see page 66)

SLUGS (see page 68)

NEMATODES (EELWORMS)

Root-lesion nematodes (see page 74)

Cereal cyst nematode (*Heterodera avenae*) (Plate 56)
Cereal cyst nematode affects winter and spring oats and maize to a greater extent than other cereals, the descending order of susceptibility to damage being spring oats, winter oats, maize, spring wheat, winter wheat, barley and rye. A run of barley or wheat crops may be only marginally affected by the pest, but a crop of oats following this run may be seriously affected by relatively low nematode infestations. Oats should therefore *not* be used as a 'break' from wheat or barley especially on light land without first making sure that nematode numbers are below danger level. This can be done by means of a pre-cropping soil test.

Damage symptoms
Both winter and spring oats can be severely stunted or even killed by the nematode. The leaves or affected plants are yellow or reddish in colour and give the appearance of suffering from drought and/or severe nitrogen deficiency. The whole crop may be affected but more usually the worst injury occurs in patches which coincide with those parts of the field where the soil is lightest.

The root systems of attacked plants are excessively branched and shallow and, from June onwards (occasionally much earlier in winter oats), while cysts are visible on the roots. At flowering, panicles of attacked plants are small and carry blind florets or undersized grain.

Life history
Infested soil can contain several millions per hectare of brown, lemon-shaped cysts about 0·1 cm long, each of which is a packet of about 400 eggs surrounded by the hardened coat of the dead female worm. The eggs and/or young larvae may remain within this protective cyst for many years, although about 60 per cent of the contents hatch every year even in the absence of a host crop.

Such larvae probably die within a few months. Most larvae emerge in the spring months in British soils. A few may emerge in autumn and invade winter-sown cereals.

Hatching rates increase in response to chemicals (root diffusates) produced by the cereal roots. The juveniles move in soil moisture films to invade both seminal and crown rootlets. This prevents further root extension and the plant responds by throwing out branch rootlets near the point of larval entry. These branch roots may in turn become infested, so that 'knots' of roots are formed and the whole root mass becomes matted.

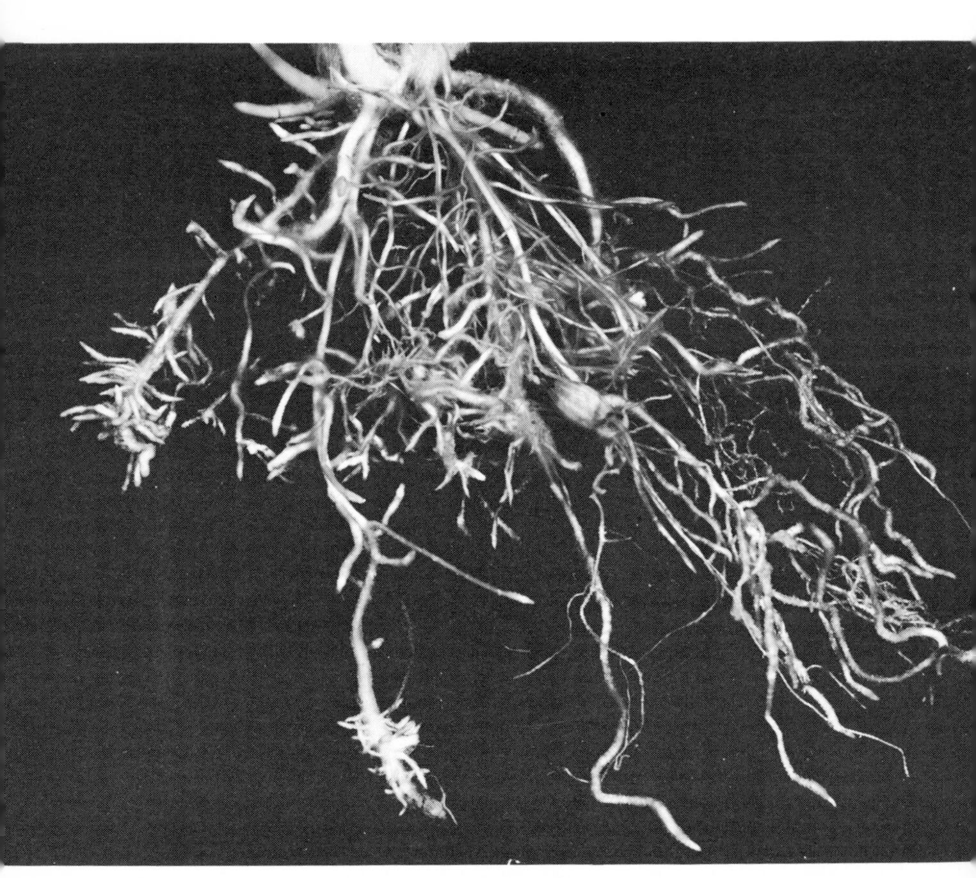

Plate 56. Cereal cyst nematode. Infected oat plant showing shallow root system with many short branching 'twiggy' rootlets.
Welsh Plant Breeding Station.

Those larvae destined to become female worms feed within 'transfer cells' formed inside the root tissue. The larvae quickly lose their eel-like form and swell into lemon-shaped females which burst through the root surface to remain attached to the root by the head end. Meanwhile, the larvae becoming male worms retain their slender shape and, when mature, escape from the roots.

After fertilisation, the female worm dies and her white skin changes to dark brown, this colour first hidden beneath a white crystalline layer. As the cereal roots rot, the cysts become free in the soil. Only one generation is completed each year.

Yield losses

Injury to cereal roots resulting from cereal cyst nematode attack is attributable to (a) root invasion by the larvae, (b) 'transfer cell' formation within the central root tissues, (c) breakage of the root cortical tissues when the males and females emerge and (d) gross changes in the form of the root system.

Yield losses have been shown to amount to 375 kg/ha for Blenda spring oats for every increase of ten eggs per gram of soil before sowing. Using soil nematicides, yield increments of 1880 kg/ha have been obtained in spring oats, although it must be emphasised that the nematicides have effects on other factors besides cereal cyst nematode.

Pathotypes

Three races or pathotypes of *H. avenae* are known to exist in Great Britain. These look alike under the microscope but differ in their ability to produce cysts on certain host plants, principally barley cultivars.

Races 1 and 2 are widespread in England and Wales, whereas Race 3 is much less common. Races 1 and 2 are distinguished by the reaction of Drost barley and some other cultivars, which are resistant to Race 1 and susceptible to Race 2.

Four pathotypes of *H. avenae* are known in the Netherlands, two in Sweden, six in Germany and two in Denmark. Some of these are identical with those present in Britain. Populations of the nematode able to reproduce on No. 191 barley have been identified in Sweden, Germany, Norway, Australia and India.

Resistance in oats is more complex than in barley (see page 167). In *Avena sterilis* the resistance is probably controlled by two dominant genes, in the cultivar Mortgage Lifter by two recessive genes, and in *A. byzantina* by a single dominant gene.

The mechanism of resistance to *H. avenae* is not fully understood. Resistant barley cultivars are often as readily invaded by larvae of the nematode as susceptible ones, but subsequent development of the nematode within resistant plants is slower. The number of adult females produced on the roots of resistant plants is very small, but adult males are produced on resistant and susceptible plants in approximately equal numbers. The situation is similar in resistant and susceptible oat cultivars, except that fewer larvae develop within the roots of resistant plants.

All British pathotypes of cereal cyst nematode are able to reproduce on grasses. A scheme showing how the pathotypes can be distinguished is given in the following table.

	British Race		
	Race 1	*Race 2*	*Race 3*
Barley			
Rika, Herta (and most common varieties)	S	S	S
Drost	R	S	—
No. 191 barley, Ansgar, Sarbarlis, Tyra	R	R	S
Morocco/Marocaine	R	R	S
Harlan 43	RS	RS	R
Regatta	R	R	—
Vista	R	R	—
Wheat			
Most varieties	S	S	—
Loros	R	R	—
Oats			
Most varieties	S	S	—
Panema	R	R	S
Avena sterilis	R	R	R
Trafalga	R	R	—
Rye			
Most varieties	S	S	—

S = susceptible
R = resistant
RS = partially resistant
— = not tested

Control
Natural control, exerted by fungal pathogens, has reduced CCN populations and serious damage is uncommon. However, as oats suffer so much more damage than other cereals, over-cropping with oats should be avoided, especially on light or medium-textured soils. Growing oats after a run of successive wheat or barley crops (presumably as a 'break' to reduce root fungal diseases) should only be done when pre-cropping soil samples show that nematode numbers are sufficiently low. An oat 'break' should be followed whenever possible by a non-host crop before returning to cereals.

Fields known to be heavily infested should not be sown to oats. Three or more successive non-host crops (e.g. roots, beans, rape, lucerne, clover) usually bring the infestation down to a safe level. Grasses in commercial use are inefficient hosts of the nematode, which falls to less than 5 per cent of its original numbers after three successive years of grass.

Some success has been obtained by using spring oats as a trap crop to depress numbers of cereal cyst nematode. The oats stimulate hatching and root invasion by larvae, but by ploughing the crop in early June the larvae are unable to complete their development.

Soil nematicides are at present far too costly to be used against cereal cyst nematode.

Greater promise lies in breeding oat cultivars which are resistant to the nematode. Most winter and spring varieties are susceptible, but wild oat I 376 *Avena sterilis*, carrying resistant genes, has been crossed with winter and spring oats in breeding programmes at the Welsh Plant Breeding Station to produce resistant cultivars. The spring oat resister cv. Nelson (of complex parentage from Sweden) is of restricted usefulness because it is very sensitive to invasion damage and consequently does not yield well on infested land. Winter oat varieties are rather more tolerant of larval invasion, and the resistant variety Panema prevents multiplication of the nematode and yields fairly well in infested soil. It is also resistant to stem nematode (*Ditylenchus dipsaci*).

Cereal root-knot nematode (see page 167)
Oats are a poor host of this nematode and root galls develop only under the high temperatures obtained in glasshouse pot experiments. It is therefore likely that under British soil conditions, both winter and spring oats can be regarded as tolerant to attacks by *M. naasi*, although differences in susceptibility between oat

genotypes have been confirmed in tests with two Welsh isolates of the nematode.

In field trials on infested land in California, resistant oat varieties have been used in cropping sequences compared with continuous barley runs and have doubled the yield of subsequent susceptible barley crops.

Stem nematode (*Ditylenchus dipsaci*) (Plate 57)

Stem nematode is an important pest in some years. It exists as several races each with a different range of host plants. The race in Britain which attacks oats also attacks mangels, sugarbeet, field beans, peas, vetch, onions, parsnip, carrot, rhubarb, teasel, turnip, rape and strawberry, as well as many weeds including chickweed, mouse-ear chickweed, black bindweed, common orache, fat hen, cleavers, scarlet pimpernel and kidney vetch. The same or possibly a different race attacks rye on the Continent, and recent information confirms that a race attacks maize there, but neither rye nor maize are commonly attacked in the British Isles.

Damage symptoms

Winter oats are more vulnerable than spring oats because the cool moist soil conditions from autumn until early spring favour nematode activity. Spring oats are invaded in the period following sowing while the soil is still cool and moist.

The oat plant shows symptoms of damage some weeks after brairding. Bases of shoots and tillers become twisted and swollen—a condition known as 'tulip root' or 'segging'. The basal tissues have a pale, spongy appearance. Such gross changes are caused by the activities of the nematodes within the plant. They produce enzymes or similar substances which dissolve the middle layer of cell walls in the stem parenchyma; soft spongy galls interspersed with air spaces appear, within which the nematodes feed and reproduce rapidly. Badly infested plants are often pale in colour and rot at ground level, so that they can be easily pulled out of the soil leaving the roots behind. Most infested plants fail to flower but those which do carry a small panicle with poorly filled grain.

'Tulip root' symptoms may be confused with those caused by frit fly (see page 216) and, to make matters more difficult, the two pests sometimes occur together. In frit fly attacks, the oat plant does not show basal swelling and distortion, only the central leaf turns yellow ('deadheart') and contains the white maggot.

Plate 57. Stem nematode. Winter oat plants showing 'tulip root' symptoms.

Life history

The adult worm is small, slender and colourless and normally cannot be seen with the unaided eye. It is not a cyst-forming species and is found fairly close to the soil surface, where it can remain alive for at least a year provided the soil remains moist. Unlike cereal cyst nematode, stem nematode is more common in heavy-textured soils with a high clay content.

Male and female worms remain eel-shaped and live as parasites in the host plant tissues. Eggs are laid inside the plant and hatch quickly to give rise to larvae which are smaller versions of the adult. Stem nematodes frequently leave the plant to move in moist soil and invade fresh hosts. When infested plants dry up or become mature the pre-adult larvae become dormant and can resist desiccation for prolonged periods before becoming reactivated under the influence of moisture. Such dormant larvae are readily transmitted in infested soil, straw, manure and plant debris.

Control

The reproductive capacity of this nematode is so great that very few are needed at the start of the season to cause trouble. Susceptible host crops should not be grown on infested soil for at least two years. Wheat and barley are immune in Great Britain and can safely be grown in place of oats.

As the pest can 'tick over' on several weed species, strict attention to weed control is vitally important.

Oat straw or mangels from infested fields should not be carted to other fields. Farmyard manure prepared from infested straw is another means whereby the pest is spread around the farm.

The resistance and/or tolerance found in several land-race types including Grey Winter and in such bred cultivars as Unique, S81, S172 and S225 has been used by plant breeders to give us resistant varieties such as Maris Quest, Peniarth, Pennal and Panema winter oats and Manod, Milford and Early Miller spring oats. These varieties, especially the last named, can suffer damage if subjected to invasion by very large numbers of stem nematode, which cause cell necrosis in the host plant tissue. Normally, resistant varieties yield satisfactorily when susceptible oats are almost a total failure.

Resistance in Grey Winter has been shown to be due to a single dominant gene. Resistant and susceptible genotypes are equally invaded by the nematode, but subsequent development and egg-laying by the nematode is much slower in resistant oat varieties, the life cycle often taking twice as long as in susceptible ones. Changes in resistant host plant tissues are much less evident following nematode invasion.

BIRDS

Apart from suffering the usual damage by various kinds of birds (see page 75), oats are particularly prone to attacks by flocks of sparrows before harvest. Ripening grains are flattened and a white deposit left on the grain coat.

MAMMALS (see page 77)

Chapter 7

OAT DISEASES

PLANT DAMAGE SYMPTOMS

Disorder	*Cause*	*Symptom*	*Page*
SEEDLINGS			
Failure to emerge	Organomercury injury	Seed does not germinate or seedlings with short thickened shoots and stunted roots.	23
	Gamma-HCH injury	Similar to and often associated with organomercury injury but shoots and roots of seedlings club-shaped.	23
Delayed emergence	Deep drilling	Seedlings yellowish and delayed, sometimes in patches.	84
'Rugby stocking'	Deep drilling	Seedlings with distinct discoloured bands on newly emerged first leaves.	84
Seedling blight	*Fusarium* spp. *Pyrenophora avenae*	Seedlings small, stunted, discoloured, may be killed before or soon after emerging.	232
Browning root rot	*Pythium* sp.	Patches of yellowed seedlings, roots brown.	231
Root rot	*Rhizoctonia solani*	Distinct patches, poor growth, sometimes red-purple discolouration.	231
Snow rot	*Typhula incarnata*	After snow patches of yellowed or rotted plants; small round brown fungal bodies on mould at plant base.	232
Frost lift	Winter frost	Small or large patches of sickly plants in spring.	86

Disorder	*Cause*	*Symptom*	*Page*
ROOTS AND STEM BASES			
Take-all	*Gaeumannomyces graminis* var. *avenae*	Roots and (later) stem bases blackened; in patches or at random.	232
Brown foot rot	*Fusarium* spp.	Stem bases brown, later with pinkish spore masses.	232
Eyespot	*Pseudocercosporella herpotrichoides*	Eyespot lesions on stem bases.	232
Sharp eyespot	*Rhizoctonia cerealis*	Lesions more clearly defined than for eyespot.	233
LEAVES AND STEMS			
Crown rust	*Puccinia coronata*	Orange-brown pustules mainly on leaves.	233
Black stem rust	*Puccinia graminis*	Orange-brown, later black, pustules mainly on stems.	234
Mildew	*Erysiphe graminis*	White fluffy pustules on leaf surface.	234
Leaf spot	*Pyrenophora avenae*	Short brown-purple stripes on lower leaves, spots on upper leaves.	235
Speckled blotch	*Leptosphaeria avenaria* (*Septoria avenae*)	Small purple-brown spots, orange margins, black spore cases on centre of spot.	236
Halo blight	*Pseudomonas coronafaciens*	Pale spots, brown margin with pale brown halo.	236
Red leaf (Barley yellow dwarf)	Barley yellow dwarf virus	Red-purple discolouration extending from tip towards base of leaf, upper leaves or all leaves affected; isolated plants at random, sometimes patches especially in winter oats.	237
Oat mosaic	Oat mosaic virus	Only in winter oats,	237

Disorder	*Cause*	*Symptom*	*Page*
		plants with yellow mottle in young leaves; plants stunted in patches.	
Oat golden stripe	Oat golden stripe virus	After earing bright yellow stripes on upper leaves; in wet patches.	238
Grey leaf	Manganese deficiency	Large grey blotches, especially in middle of leaf.	238
EARS			
Smuts	*Ustilago hordei* *Ustilago avenae*	All grains replaced by a mass of black spores which are shed, leaving a bare stem.	239
Ergot	*Claviceps purpurea*	Hard black fungal bodies replace grain in a few spikelets; uncommon.	240
Scab	*Gibberella zeae*	Red fungus with black spore cases embedded; on surface of spikelets.	144
Black mould	*Cladosporium* spp.	'Sooty' black mould on surface of grains and spikelets.	145
'Blast'	Several causes	Spikelets, especially lower ones, bleached, empty.	240
'Blindness'	Frost damage	Spikelets bleached and blind in groups mainly at top of panicle.	240
Copper deficiency		Terminal spikelets blind.	240
Distortion	Herbicide injury	Ears malformed.	240

Browning root rot (*Pythium* spp.)
This root rot (see page 85) has been recorded very occasionally on oats in Britain but has attracted little attention from pathologists. It is not associated with any severe crop damage and its importance in oat production is unknown.

Root rot (*Rhizoctonia solani*)
Outbreaks of this disease on oats are reported very occasionally. The disease is described in the section on barley (page 173).

Snow rot (*Typhula incarnata*)
This is noticed on plants when the snow melts away (see under barley page 174).

Take-all (*Gaeumannomyces graminis* var. *avenae*, syn. *Ophiobolus graminis* var. *avenae*)
Oats are almost immune to the take-all fungus commonly occurring on wheat and barley and are sometimes used as a 'break' crop in connection with this disease. However, they are susceptible to a distinct strain of the fungus, *G. graminis* var. *avenae*, which also attacks wheat and barley. This strain occurs most commonly in western and northern parts of Britain where oats are grown more frequently, although occasional outbreaks have occurred elsewhere. The effects of the disease on oat crops are much the same as those described for wheat (see page 87) but since oats are not usually grown intensively (partly because of eelworm attacks, page 221 and page 225), severe and extensive attacks are rarely seen.

The oat strain of *G. graminis* is the one most commonly recorded on pasture and turf grasses and it is rather surprising that it is not more widespread in cereals, especially following ploughed-out grass.

Seedling blight, brown foot rot, ear blight and **scab** (*Fusarium* spp.)
Although it is generally agreed that wheat (page 95) and barley are more susceptible to attack by these fungi, oats and rye are also attacked and exhibit the same range of diseases, from seedling blight to foot rot and 'whiteheads' and subsequently the ear blight stage. Both the seedling blight and foot rot phases can be quite severe in oats, while the ear blight phase is recorded only rarely and then almost invariably in wet seasons. Oats are also susceptible to scab (see page 144) and slight attacks are occasionally reported, chiefly in western and northern areas.

Eyespot (*Pseudocercosporella herpotrichoides*)
Although eyespot is occasionally reported on oats, this crop is generally considered to be much less susceptible to infection than either wheat (see page 98) or barley. However, in Ireland eyespot has occasionally caused severe damage. Oats are thought to serve as a means of carry-over between wheat crops so that, for example, a sequence of wheat, roots, oats, beans, wheat is likely to result in higher levels of infection in the second wheat crop

than a sequence of wheat, roots, beans, wheat, which provides a complete two-year break.

Sharp eyespot *(Rhizoctonia cerealis)*
Oats are considered to be more susceptible to sharp eyespot than is wheat (page 106) but severe attacks are rarely reported. Oats occupy a very small area compared with wheat so it is not surprising that field records of the disease are relatively infrequent and the effects of the disease, in terms of crop losses, have not been assessed.

Crown rust (*Puccinia coronata*)
Crown rust is the common rust disease of oats but does not occur on any other cereal. It affects several grasses, notably perennial rye-grass, but the strains that affect grasses do not affect oats. The alternate hosts are buckthorn (*Rhamnus catharticus*) and alder buckthorn (*Frangula alnus*) on which the cluster-cup (aecidial) stage is produced.

The first sign of crown rust on oats is the appearance of orange pustules containing uredsospores scattered at random mainly on the leaf blade. The sheath and more rarely the stem and panicle may also be affected. The disease is spread by the wind-borne uredospores until later in the season when black pustules, containing teliospores, are produced. The teliospore, seen under the microscope, has crown-like appendages at its apex and it is this feature that gives the disease its name. Teliospores remain dormant on straw until spring when they germinate to produce basidiospores which can only infect the alternate host on which the cluster cups are formed. Aecidiospores from the cluster-cups then infect oats to complete the life cycle. This aecidial stage is thought to play an insignificant role in the survival and spread of the disease.

Crown rust survives the winter as uredospore pustules or mycelium in the leaves of volunteer plants and winter oats but in some parts of the country such as Scotland, the fungus does not survive in some years. Buckthorn may be the source of some local outbreaks in the spring but the main source is winter oats, especially in the milder parts of the country. The disease spreads from these over-wintering crops to spring-sown ones by means of wind-blown uredospores.

Crown rust is favoured by warm, wet and humid weather. In most seasons attacks occur relatively late and yields are not seriously affected. However, when attacks are severe during the period from panicle emergence to the milky ripe stage the effect on yield is serious, the grain being light and shrivelled. In areas where the disease is consistently a problem the best control measure available is the use of resistant varieties. Late tillers and volunteers should, as far as possible, be destroyed before the winter crop emerges and spring crops should not be sited adjacent to winter crops. Some fungicides which give an efficient control of other cereal rusts also control crown rust but little work has been done on this subject.

Black stem rust (*Puccinia graminis*)
The orange and, later, black pustules are more prominent on the stem. The disease is rare in England (see under wheat page 116).

Mildew (*Erysiphe graminis*)
Mildew is a common disease of oats and can be very damaging. The area of oats has become relatively small in recent years and severe attacks of mildew are uncommon but intensification of the crop under modern farming conditions would be conducive to more frequent attacks. A detailed account of cereal powdery mildew is given for wheat (page 118) and for barley (page 181).

The white pustules of the mildew fungus are most common in the leaf blade. The sheath may also be affected but not the panicle. When conditions are not favourable for mildew development, infections show as small brown-black spots with a red or purple discolouration around them. The typical superficial white fungal growth is very much reduced and in some cases cannot be detected with the naked eye but needs a microscope or at least a good hand lens.

The life history of the fungus is similar to that in wheat but late green tillers, which occur more commonly in oats, are said to be relatively more important in carrying the fungus through the harvest period. The epidemic in oats has not been observed in as much detail as the epidemics in wheat and barley but the factors affecting disease development are probably similar; those in winter oats being similar to the ones in winter wheat (page 118), and in spring oats to those in spring barley (page 181).

As with the other cereal mildews, yield losses are related mainly to the earliness and severity of mildew attack. An estimate of the percentage yield loss for oats is obtained from the formula $2{\cdot}5\sqrt{M}$ where M is the percentage leaf area affected by mildew at the complete ear-emergence growth-stage. This probably underestimates the damage especially from early attacks. Average grain losses in affected crops are about 5–10 per cent but losses as high as 40 per cent have been recorded.

Mildew in oats may be controlled by using resistant varieties (see NIAB Recommended Varieties) or with fungicides. Some seed treatments and sprays used for barley mildew control are also effective against oat mildew. Sprays should be timed as recommended for barley (see page 186).

Leaf spot and **seedling blight** (*Pyrenophora avenae* [*Drechslera avenae*, syn. *Helminthosporium avenae*])

This disease was at one time regarded as an important disease of oats. It was well controlled by organo-mercury seed disinfectants but in the 1960s strains resistant to organo-mercury preparations emerged and these strains became dominant. Although the disease remained common and often was not controlled by organo-mercury seed disinfectants, no cases of severe disease outbreaks were reported. No reasons for this were identified and in terms of its effects on yield the disease become unimportant. In any case leaf spot can now be well controlled by some of the newer systemic seed treatments.

The fungus is carried on the seed as spores or in the husk and seed coat as dormant mycelium. When the seed is sown the fungus spreads to the sheath (coleoptile) as it emerges from the seed. In the primary phase of the disease the fungus passes from the sheath to the first leaf as it emerges and then in a similar manner to the next two or three leaves. The first three or four leaves have short stripes which are brown, often with purple margins. Subsequent leaves are healthy as they unfold but may be infected later, in the secondary phase of the disease, by spores produced on the stripes of the leaves infected in the primary phase. In the secondary phase only a few spots occur and these are red-brown with purple margins. Spores produced on these spots infect or contaminate the developing grain.

As mentioned above, the present status of the leaf spot fungus as a serious pathogen is in some doubt. However, earlier work showed that serious losses were associated with the primary phase of the disease through death of seedlings before or soon after emergence (seedling blight). Severely affected seedlings which survived, tillered poorly and produced fewer spikelets, some of which were 'blasted' and yielded poorly. The secondary phase had little effect on yield but ensured that the fungus reached the developing grain. The primary phase is more serious at low temperatures and the secondary phase is favoured by wet or humid conditions.

Control

The most important source of infection is the seed. The common strains are resistant to organomercury disinfectants but effective alternatives are now available.

Infected oat debris from a previous crop is though to be unimportant as a source of the fungus and the perfect stage (*Pyrenophora*) has been found only in Scotland and rarely.

The more severe effects of seedling blight may be avoided by late drilling of spring oats when soils are warmer. However this has its danger in that late sowing itself may lead to low yields and the late-sown crop is more liable to suffer severe frit fly damage (page 216).

In practice there is no need to take routine precautions against this disease, except for the use of an effective seed disinfectant.

Speckled blotch (*Leptosphaeria avenaria* [*Septoria avenae*])

This is a common and occasionally damaging disease. The spots on the younger leaves are oval, about 0·5–1·5 cm across, dark purple-brown with an orange margin. In the later stages of infection the lesions are commonly found at the base of the leaf blade at its junction with the sheath. The affected areas near black-brown dots, the spore-producing pycnidia, which impart the speckled appearance to the blotches. Later in the season the fungus at the leaf base may attack the stem which becomes brown-black and weakened so that it breaks in the diseased area. A spotting of the panicles and glumes also occurs and the fungus can infect the seed. Like other *Septoria* diseases, speckled blotch is favoured by wet and humid weather.

The main sources of the disease are infected straw and seed. The perfect stage, which is believed to have been found on stubble in Scotland, may also be a source. This is one of the diseases, at present relatively unimportant, which may assume greater significance if the present restricted oat area is increased.

Halo blight (*Pseudomonas coronafaciens*)

Diseases caused by bacteria are uncommon in Britain. This bacterial disease, though apparently fairly common, is not economically important. The bacteria are seed-borne and spread from infected seed to the first leaves and then by wind and rain to later-formed leaves and to the panicles. The spots have a brown dead centre surrounded by a pale green (later grey-brown) halo about 1·5 cm long. Spots may coalesce to give irregularly shaped diseased areas which may be confused with grey speck (see page 238).

At present there is no need to apply specific control measures. Seed treatments do not control the disease. Warm-water treatment and antibiotics have been effective in eradicating bacteria in seed, so that control measures are available if they ever become necessary.

Red leaf or **barley yellow dwarf** (Barley yellow dwarf virus, BYDV)

This is a fairly common disease in oats and for many years before the virus was recognised as the cause, the symptoms, referred to in the literature as 'red leaf', were frequently attributed to aphid injury. In affected plants the leaf is bright purple-red from the tip extending towards the base. The upper leaves or all of the leaves may be affected, depending on the time of infection, and affected plants occur immediately adjacent to, and mixed with, normal green plants and this distinguishes the disease from many other diseases and disorders. Occasionally severe stunting in large patches has occurred in crops of winter oats especially in Wales and south-west England when crops are sown in early autumn and particularly when they are sown immediately after ploughing grass.

Apart from the symptoms, the general account of the disease in barley (see page 200) also applies to oats.

Oat mosaic (Oat mosaic virus)

This disease, which appears to be the same as the oat mosaic known for many years in North America, has been noticed in a few isolated crops of winter oats in western and southern parts of England and in Wales and can cause serious yield losses. Symptoms on winter oats are seen in early spring when affected plants occur in distinct small or large patches. The symptoms include a yellow-green mosaic which is especially noticeable on the youngest leaves. Affected plants, except those around the edges of the patch, are very stunted and loss in yield can be severe. Spring oat varieties can be infected but are symptomless. This is partly because they are exposed to infection for a shorter time and partly because symptoms are expressed at fairly low temperatures such as occur in winter and early spring.

The virus is soil-borne and there is some limited evidence that it is transmitted to oat plants by the soil fungus *Polymyxa graminis* which is also associated with barley yellow mosaic virus (page 205). The virus is not seed-borne and is apparently restricted to *Avena* species. Some varieties are tolerant, that is they show mild or no symptoms and their yields are much less affected. Such varieties provide the only practical means of control where it is necessary to grow oats on a known infected site.

Oat golden stripe (Oat golden stripe virus)
This rare disease occurs only in association with oat mosaic and has been recorded in only a very few of the small number of oat mosaic outbreaks recorded. Plants with typical symptoms of oat mosaic (see above) earlier in the season develop bright yellow stripes on the flag and second leaves just after earing. Affected plants occur in patches, often in poorly drained areas. Because the disease has only been found in association with oat mosaic, it is difficult to ascribe a degree of damage specifically to it. However, it is unlikely to do other than reduce further the poor yield caused by oat mosaic so it is fortunate that it occurs so infrequently. It seems likely that oat golden stripe virus is transmitted by a soil-inhabiting fungus but this has not been proven.

Grey speck or **grey leaf** (manganese deficiency)
Manganese deficiency occurs on a range of soil types but is most common on peaty and organic soils. It is also common, although generally less severe, on sandy soils. It is often associated with alkaline soils but may occur on soils with a lower pH. It is particularly associated with soils that have been recently limed. All the cereals are affected and the descending order of susceptibility is oats, wheat, barley and rye.

In oats, affected plants tend to occur in patches of variable size or at random in ill-defined areas. The plants become pale green and then develop pale green spots, often in the middle part of the leaf, which become grey, sometimes with purple-brown margins. The spots are fairly large and coalesce to form irregular areas and streaks. The leaf often collapses in the affected middle areas so that the upper portion, which may be still green, hangs down. These symptoms occur in early summer when the plants are about 15–20 cm high and the affected areas in the crop appear badly scorched. Later in the season there may be some recovery, especially if the soil is moist, and the upper leaves then remain green. Affected plants have poorly developed root systems and in summer droughts may die.

Manganese deficiency can be corrected by spraying the affected crop at the first sign of symptoms with 6–9 kg per hectare of manganese sulphate plus wetter in at least 250 litres of water. Treated crops show signs of recovery in 10–14 days. The treatment may cause leaf scorch if crops under moisture stress are sprayed in bright sunshine and especially if more concentrated sprays are used. On fields when manganese deficiency is expected, the spray can be applied when plants are 10–15 cm high and before symptoms are seen. Where the deficiency is severe it may be necessary to spray more than once.

Smuts (*Ustilago hordei* and *Ustilago avenae*)
Although oats, like the other cereals, are attacked by two species of smut fungi, *U. hordei* causing covered smut and *U. avenae* causing loose smut, in practice the two diseases (and the two fungi) are difficult to distinguish from each other. For practical purposes, they can be regarded as one disease. It is fortunate that they both behave as seed contaminants, like the covered smuts of wheat and barley, and both are effectively controlled by seed disinfectants.

Although the loose smut spores, and some of those of the covered smut, are blown about in the crop at flowering time, it seems that they do not penetrate as far as the embryo as they do in barley or wheat loose smut, but remain either as spores lodged between the outer husks covering the grain or as dormant mycelium in the outer parts of the grain. The precise significance of dormant mycelium is a matter for conjecture but is of academic interest only, since if it can act as a source of seedling infection, it does so in the same manner as do the spores, that is, it commences growth at about the same time as the seed and infects the very young seedlings soon after germination.

Thereafter, the pattern of development closely parallels that of wheat bunt (see page 136), except that under good growing conditions, the rate of growth of the plant may outstrip that of the fungus, so that at harvest time the main tiller of an infected plant may be free from the disease while the later secondary tillers show smutted ears.

Since the general adoption of seed disinfection with organomercury these diseases have become comparatively rare in Britain. The newer systemic seed treatments also provide an effective control.

Ergot (*Claviceps purpurea*)
Ergots are hard, black fungal bodies which replace the grain in the panicle. The disease is rare in oats but more common in other cereals (see page 140).

Blast
This describes a condition rather than a specific disease, in which some of the spikelets, especially the basal ones, at the time of emergence of the panicle are white and shrivelled and produce no grain. This can be due to any one or more of the several factors which lower the vitality of the plant at a critical time in spikelet development before the panicle emerges. Factors which have been implicated include water shortage, poor or excessive light, faulty nutrition and sudden defoliation such as may be caused by pest attack. Factors which operate over a long period of time are not likely to cause the condition.

The term 'blast' sometimes includes the occurrence of sterile spikelets in a panicle but in contrast to the condition described above this may be due to factors operating before and/or after the panicle emerges. Such factors include severe attacks of leaf diseases, barley yellow dwarf virus, soil-borne disease, adverse soil conditions, frost damage and some pests such as frit fly and oat spiral mite.

Late frost damage
Frost can damage leaves and flowers leading to sterile spikelets as mentioned above under 'blast'. More detailed descriptions of the kinds of damage caused by frost are given under wheat (pages 86 and 148).

Copper deficiency
Copper deficiency is noticed less frequently in oats than in wheat (page 149) and barley but this is probably only because oats are grown less frequently on copper-deficient soils. In south-east Scotland a condition known as 'wither tip', in which the terminal spikelets are blind, is caused by copper deficiency.

Herbicide injury
Growth regulatory herbicides applied earlier or later than recommended can cause serious damage to cereals (see barley, page 211). In oats, applications earlier than recommended cause tubular leaves and distortions in the form and arrangements of the spikelets. The head often has an upright, stiff, bunched

appearance. The spikelets are bunched together, arranged in whorls or bunched at the ends of branches. The spikelets may be fused, distorted or blind. The panicle may be split or trapped in the leaf sheath causing the stem to buckle.

Spraying later than recommended does not cause distortion but may result in blindness or shrivelled grain and is more likely to cause serious loss of yield than is spraying earlier than recommended.

Chapter 8

RYE PESTS

PLANT DAMAGE SYMPTOMS

a. *Seed or seedling damaged at or before emergence*

Symptom	Cause
Seeds missing from drill row....................	Birds (p. 75) Rats, mice (p. 77)
or hollowed out	Slugs (p. 68) Wireworms (p. 55) Mice (p. 77)

b. *Shoot develops but does not reach soil surface*

Symptom	Cause
Shoot short and swollen, root tips club-shaped	Seed treatment injury (p. 23)
Shoot brown and shown signs of feeding....	Slugs (p. 68) Leatherjackets (p. 162) Wireworms (p. 55) Wheat bulb fly larvae (p. 246) Frit fly larvae (p. 246)
Shoot long and often twisted	Deep sowing (p. 84) Soil capping (p. 86)

c. *Seedling or tillering plant damaged*

i. Whole plants affected, usually yellow first

Symptom	Cause
Plants pulled out and left on soil	Birds (p. 75) Rats/mice (p. 77)
Plants bitten near soil level................	Slugs (p. 68) Leatherjackets (p. 162) Wireworms (p. 55) Grass moth caterpillars (p. 49)

Symptom	Cause
	Swift moth caterpillars (p. 49) Cutworms (p. 49) Chafer grubs (p. 57) Wheat shoot beetle larvae (p. 66) Millepedes (p. 66)
Plants not bitten...............................	Frost damage (p. 86) Deep sowing (p. 84)
Shoot swollen or mis-shapen...............	Gout fly larvae (p. 157) Stem nematode (p. 247)
Plants slightly stunted, leaves spotted or flecked...	Aphids (p. 246) Thrips (p. 65)
ii. Centre shoot yellows and dies (deadheart), outer leaves stay green..	Wheat bulb fly larvae (p. 246) Frit fly larvae (p. 246) Grass moth caterpillars (p. 49) Common rustic moth caterpillars (p. 54) Grass and cereal fly larvae (p. 43) Yellow cereal fly larvae (p. 43) Bean seed fly larvae (p. 42) Wireworms (p. 55)
iii. As above, but a small neat hole often present near shoot base	Flea beetles (p. 59) Wheat shoot beetle larvae (p. 58)
iv. Leaves bitten often well above ground	
Long narrow holes.........................	Slugs (p. 68)
Roughly circular holes	Grass moth caterpillars (p. 49)
v. Leaves neatly cut off	
Horizontal bitten edge, tips missing...	Mammals (p. 77)

V-shaped leaf bite, tips often lying on soil..	Birds (p. 75)
vi. Leaves torn, with ragged ends............	Slugs (p. 68) Leatherjackets (p. 162)

d. *Plant damaged after tillering stage*

i. Whole plant sickly, yellow or reddish Root system with small thickened galls.	Cereal root-knot nematode (p. 247)
ii. Some shoots yellow, others healthy	Grass moth caterpillars (p. 49) Common rustic moth caterpillars (p. 54)
iii. Some shoots swollen, others healthy	Gout fly larvae (p. 157) Stem nematode (p. 247)
iv. Stem surface near upper nodes pitted with saddle-shaped depressions.........	Saddle gall midge larvae (p. 47)
v. Leaves bitten or discoloured Long narrow strips eaten from leaf blade	Slugs (p. 68) Cereal leaf beetle (p. 219) Barley flea beetle (p. 164)
Irregular areas bitten from leaf edge	Grass moth caterpillars (p. 49)
Discoloured blotches on leaf surface.....	Aphids (p. 246)
Silvery marks on leaf surface	Thrips (p. 65)
Leaves with blister mines	Cereal leaf miner larvae (p. 45)
vi. Shoots bend or break and fall over Shoot bent above ground level, brown seedlike bodies beneath leaf sheath near cut edge..............................	Hessian fly (p. 161)
Shoot cut off with diagonal bite or peck marks, heads often stripped............	Birds (p. 75) Mammals (p. 77)

Symptom	Cause
Shoot breaks off near upper nodes.......	Saddle gall midge larvae (p. 47)

e. *Damage to flowering heads*

Symptom	Cause
Distal ears white, distorted or shrivelled....	Frit fly larvae (p. 246)
Whole ear white or aborted....................	Hessian fly larvae (p. 161)
Ears distorted, stems often twisted	Wheat gall nematode (p. 247) Frit fly larvae (p. 246)
Ear often one-sided, groove runs down side of stem...-	Gout fly larvae (p. 157)
Ears sticky, often covered with extruded fluid..-	Aphids (p. 246)

f. *Damage to ripening grains*

Symptom	Cause
Grains missing.....................................	Lemon wheat blossom midge (p. 46)
Grain shrivelled or incompletely developed.	Grain aphid (p. 246) Orange wheat blossom midge (p. 46) Thrips (p. 65) Birds (p. 75)
Grain blackened and/or partly eaten	Rustic shoulder knot moth caterpillar (p. 54)
Grain replaced by black powdery mass	Wheat gall nematode (p. 247)
Grain flattened, white deposit on outside ...	Birds (p. 75)

In terms of the range of pests which attack it, rye stands in an intermediate position between wheat and barley on the one hand and oats on the other. Thus wheat bulb fly, gout fly, Hessian fly and cereal root knot nematode readily attack wheat, barley and rye but not oats, while stem nematode attacks oats and rye but not wheat or barley.

Winter and spring rye varieties grown for grain seem equally at risk to pest damage as those varieties grown for grazing. The fact that attacks on rye are rarely encountered in the British Isles is probably due to the relatively small area of the cereal grown here rather than to its immunity or tolerance to pest damage.

INSECTS

Wheat bulb fly (see page 36)
Rye, like barley, is an inefficient host of wheat bulb fly larvae, the maggots entering rye shoots but rarely surviving to maturity.

Bean seed flies (see page 42)

Grass and cereal flies (see page 43)

Yellow cereal fly *(Opomyza florum)* (see page 43)

Gout fly (see page 157)

Frit fly (see page 216)
Autumn-sown rye is vulnerable to frit fly larval damage when the crop follows grass or a grass-infested stubble. Attacks can result in serious thinning of rye crops.

Leaf miners (see page 45)

Wheat blossom midges (see page 46)
Both lemon and orange blossom midge larvae infest the flowering heads of rye.

Saddle gall midge (see page 47)

Hessian fly (see page 161)

Leatherjackets (see page 52)

Moth caterpillars
The range of species which attack wheat (page 49) also attack rye.

Beetles (see page 55)
Those beetles and beetle larvae attacking wheat and barley similarly affect rye.

Aphids and leafhoppers
The same aphid species found on wheat, oats and barley (page 60) also colonise rye, but the effects of direct injury or virus infection or grain yield or quality of rye have not been assessed.

Thrips (see page 65)

MITES

Neither the oat spiral mite (page 220) nor the grass and cereal mite (page 66) attack rye in Britain.

MILLEPEDES (see page 66)

SLUGS (see page 68)

NEMATODES (see page 71)

Wheat gall nematode
Rye suffers damage in a similar manner to wheat (see page 71) but attacks in this country have not been recorded in recent years.

Root-lesion nematodes (see page 74).

Cereal cyst nematode (see page 221)
Both winter and spring rye root systems are invaded by larvae of *Heterodera avenae*, but the plants show little sign of damage and few cysts develop on the roots. In experimental pot tests Rheidol winter rye sown in spring behaved rather differently in that moderate numbers of cysts have been produced on its roots.

Cereal root-knot nematode (see page 167)
Although galls are readily formed on the roots of winter and spring rye, the plants seem unaffected as far as yield loss is concerned. Where patches occur in fields infested with the nematode, it is likely that other factors such as drainage or fertiliser deficiency are primarily concerned.

Stem nematode
A race of stem nematode attacks rye and produces 'tulip root' symptoms similar to those seen in oats (page 226). Such damage has been frequently recorded in the Netherlands and other countries where a substantial acreage of rye is grown each year. The oat race of stem nematode may also affect rye in the Netherlands where the variety Heertvelder has resistance derived from cv. Ottersumse.

In Britain, records of stem nematode attacking rye are very

scanty and it is not certain whether or not the oat race is responsible.

BIRDS (see page 75)

MAMMALS (see page 77)

Chapter 9

RYE DISEASES

PLANT DAMAGE SYMPTOMS

Disorder	*Cause*	*Symptom*	*Page*
SEEDLINGS			
Seedling blight	*Fusarium* spp.	Seedlings small, stunted, discoloured; may be killed before or soon after emergence.	250
Frost lift	Winter frost	Small or large patches of sickly plants in spring.	86
ROOTS AND STEM BASES			
Take-all	*Gaeumannomyces graminis*	Roots attacked, plants stunted, in patches or at random.	250
Brown foot rot	*Fusarium* spp.	Stem bases brown, later with pinkish spore masses.	250
Eyespot	*Pseudocercosporella herpotrichoides*	Eyespot lesions on stem bases.	251
Sharp eyespot	*Rhizoctonia cerealis*	Lesions more sharply defined, higher up stem than eyespot.	251
LEAVES AND STEMS			
Yellow rust	*Puccinia striiformis*	Orange/yellow pustules arranges in lines.	251
Brown rust	*Puccinia recondita*	Brown pustules scattered at random on leaf.	251
Black stem rust	*Puccinia graminis*	Orange-brown (later black) pustules mainly on stem and sheath.	251
Stripe smut	*Urocystis occulta*	Black spore masses in stripes.	251

Disorder	*Cause*	*Symptom*	*Page*
Mildew	*Erysiphe graminis*	White fluffy pustules on leaf surface.	252
Leaf blotch	*Rhynchosporium secalis*	Large grey-brown spots especially on lower leaves.	252
Leaf spot	*Septoria secalis*	Pale brown spots.	252
EARS			
Bunt	*Tilletia caries*	Inside of grain replaced by black spores.	252
Ergot	*Claviceps purpurea*	Hard black bodies replacing grains and protruding from some spikelets.	252
Ear blight	*Fusarium* spp.	Pink-red spore masses on surface of ear.	250
Scab	*Gibberalla zeae.*	Pink-red fungus with black bodies (perithecia) embedded on surface of ear.	250
Black mould	*Cladosporium* spp.	Sooty black moulds on surface; ears often thin.	145
Whiteheads	*Gaeumannomyces graminis, Pseudocercosporella herpotrichoides,*	Ears bleached, prematurely ripe, grain shrivelled. A symptom of many disorders.	250
	Rhizoctonia cerealis,		251
	Fusarium spp., etc.		250

Take-all (*Gaeumannomyces graminis* syn. *Ophiobolus graminis*)
Rye is less susceptible than other cereals to attacks by the take-all fungus but severe attacks have been noted occasionally. It is not a common cause of loss in yield. A full account of the disease in wheat is on page 87.

Seedling blight, brown foot rot, ear blight and **scab** (*Fusarium* spp.)
Rye is susceptible to attack by the brown foot rot fungi, and the seedling blight phase of the attack can be quite damaging in the early part of the season. The ear blight phase and scab are not nearly so often reported on rye as on wheat, but this is probably a direct reflection of the relatively small acreage of rye grown in Britain. Several species of *Fusarium* are involved and the diseases

they cause are described more fully under wheat (pages 95 and 144).

Eyespot (*Pseudocercosporella herpotrichoides*)
Eyespot is occasionally severe in rye and can cause extensive lodging. There are two forms of the eyespot fungus. The W-type affects wheat and barley significantly more severely than it affects rye. The R-type is equally damaging to wheat, barley and rye. The latter was uncommon and usually associated only with fields which had grown rye. However during the 1980s the R-type has become the type most commonly isolated from all cereals. This has been associated with the development of MBC-resistance in the eyespot fungus.

The account of the disease in wheat (page 98) generally applies to rye.

Sharp eyespot (*Rhizoctonia cerealis*)
In pot experiments, rye has been shown to be the most susceptible of all cereals to infection by the sharp eyespot fungus. Although the crop is often grown on sandy soils of low pH liable to drought, precisely the situation in which the disease might be expected to be severe, field records show that it has not proved troublesome so far. Symptoms and behaviour are similar to those described in detail for wheat (page 106).

Yellow rust (*Puccinia striiformis*)
This is only rarely found on rye (see under wheat page 109).

Brown rust (*Puccinia recondita*, syn. *P. dispersa*)
The pustules are dark brown and are scattered at random on the leaves. The disease may occasionally develop late in the season but it is not usually severe. Rye strains of the fungus do not affect other cereals (see wheat brown rust, page 114).

Black stem rust (*Puccinia graminis*)
This rust is not common and occurs too late to cause damage (see page 116).

Stripe smut (*Urocystis occulta*)
This smut is specific to rye and produces black spore masses in stripes on the stem and leaves. It is unusual among the smuts of temperate cereals in being soil-borne as well as seed-borne. Infected plants are stunted, darker green than usual and ears may

be attacked before emergence. The disease is uncommon but isolated severe attacks have occurred. Infection of the seedling takes place from spores on the seed or surviving in the soil.

The disease can be controlled by seed disinfection coupled with an adequate rotation.

Mildew (*Erysiphe graminis*)

The white pustules of mildew are common on leaves but the disease is less important than on wheat (page 118) and barley (page 181).

Leaf blotch (*Rhynchosporium secalis*)

The large grey-brown spots, often without the distinct margins characteristic of the disease in barley, are common on the lower leaves but the disease is not usually as severe on rye as it is on barley (page 194).

Leaf spot (*Septoria secalis*)

This disease shows as pale brown spots similar to those caused by *Rhynchosporium* and is also favoured by wet weather. The spots are usually confined to the lower leaves and sometimes bear brown spore cases (pycnidia). The disease is rarely damaging and no control measures are advocated.

Bunt (*Tilletia caries*)

This disease, in which black oily spores fill the inside of the grain, is rare. It is caused by rye-specific strains of the fungus that causes bunt in wheat (for full account see page 136) and can be controlled by organo-mercury and some other seed disinfectants.

Ergot (*Claviceps purpurea*) (Plates 44, 45, page 141)

Ergot is a common disease of grasses and cereals. The causal fungus only attacks the ear, replacing the grains by a hard purple-black body known as an ergot. In rye the ergot is long (up to 1·80 cm), slightly curved, horn-like and protrudes from the affected spikelet so that it is very obvious in the standing crop.

Rye is the most susceptible cereal, partly because its flowers remain open for long periods when infection can occur. The disease is fully described on wheat (page 140).

Chapter 10

MAIZE PESTS

PLANT DAMAGE SYMPTOMS

a. *Seed or seedling damaged at or before emergence*

Symptom	Pest
Seeds missing from drill row....................	Birds (p. 257) Rats, mice (p. 77)
or hollowed out	Slugs (p. 68) Wireworms (p. 256) Mice (p. 77)

b. *Shoot develops but does not reach soil surface*

Symptom	Pest
Shoot brown and shows signs of feeding	Slugs (p. 68) Leatherjackets (p. 162) Wireworms (p. 256) Frit fly larvae (p. 255)

c. *Seedling or tillering plant damaged*

i. Whole plants affected, usually yellow first

Symptom	Pest
Plants pulled out and left on soil	Birds (p. 257) Rats, mice (p. 77)
Plants bitten near soil level.................	Slugs (p. 68) Leatherjackets (p. 162) Wireworms (p. 256) Swift moth caterpillars (p. 49) Cutworms (p. 49) Chafer grubs (p. 57)
Shoot swollen or mis-shapen...............	Stem nematode (p. 257)
Plants slightly stunted, leaves spotted or flecked...	Aphids (p. 256) Thrips (p. 256)

Symptom	Cause
ii. Centre shoot yellows and dies (deadheart), outer leaves stay green..	Frit fly larvae (p. 255) Rustic moth caterpillars (p. 256)
iii. As above, but a small neat hole often present near shoot base	Flea beetles (p. 59)
iv. Leaves bitten, often well above ground	
Long narrow holes.........................	Slugs (p. 68)
v. Leaves neatly cut off	
Hortizontal bitten edge, tips missing.....	Mammals (p. 77)
V-shaped leaf bite, tips often lying on soil...	Birds (p. 257)
d. *Plant damaged after tillering stage*	
i. Some shoots yellow, others healthy	Rustic moth caterpillars (p. 256)
ii. Some shoots swollen, others healthy....	Stem nematode (p. 256)
iii. Leaves bitten or discoloured	
Long narrow strips eaten from leaf blade	Slugs (p. 68)
Silvery marks on leaf surface.............	Thrips (p. 256)
Whitish marks on leaf surface, webs often present...............................	Two-spotted spider mite (p. 256)
iv. Shoots bend or break and fall over	
Shoot cut off with diagonal bite or peck mark, heads often stripped.............	Birds (p. 257) Mammals (p. 77)
e. *Damage to ripening grains*	
Grains missing....................................	Birds (p. 257)
Grains shrivelled or incompletely developed	Birds (p. 257)
Grains flattened, white deposit on outside..	Birds (p. 257)

Maize differs from other cereals in being attacked by relatively few pests and was considered as a possible 'break' crop in those parts of southern England where cereal cyst nematode is a threat to intensive wheat, barley or oat growing until it was realised that maize is particularly susceptible to CCN damage. Many of the 'ley pests' which damage wheat, barley, oats and rye when these crops follow grass or grassy stubble have no effect upon grain or fodder maize.

At present, the area of maize is small and limited to counties in southern and south-east England, where it suffers serious damage by birds and frit fly. Other pests may become important if and when the area increases but so far we have not seen any significant changes on a scale comparable with those in France, West Germany, Austria and other European countries.

INSECTS

Frit fly (*Oscinella frit*)
This is the most important insect pest of grain and fodder maize (and sweet corn) in Britain, because the optimum time of sowing under our conditions leaves the seedling plant at a susceptible stage during the peak period of egg-laying by frit flies.

As in oats, the eggs are laid beneath the leaf sheath, under the coleoptile or on the soil very close to the maize plant. The maggots bore into the shoot at a point about 1·3 cm above the soil, and several grubs may feed together on the central tissues, boring in a spiral track down to the growing point. The central shoot yellows and dies ('deadheart') and the whole plant may die at this stage. Stronger plants produce fresh basal tillers which may in turn be attacked so that the whole plant has a grasslike appearance.

Occasionally, the grubs travel upwards to the shoot tip, when growth is stunted and the central tissues are brown and covered with insect frass; the shoot in this case is not killed.

Larval feeding at the growing point may result in twisting and distortion of the first six leaves, each of which may be ragged and cut with long lacerations. The leaf tips remain rolled together. Sometimes the leaves show a series of neat holes—the result of larval tunnelling within the rolled leaf.

The maize plant remains susceptible until it has passed the six-leaf stage. A survey in Essex showed that crops not treated with an insecticide at sowing sustained over 20 per cent damaged shoots, compared with only 4 per cent in treated crops. Yields of maize, particularly of crops grown for grain, are materially affected when 20 per cent or more of the primary shoots are destroyed.

Chemical control measures should be applied as a routine insurance against frit fly attacks. Insecticide granules can be applied along the drill at sowing by means of a microband granule applicator or a seed treatment used. Alternatively, an insecticide spray or granules can be applied at full crop emergence.

Leatherjackets (see page 162)

Moth caterpillars
Caterpillars of the common rustic moth (p. 54) and rosy rustic moth (p. 54) have been found attacking maize in southern England.

Wireworms (see page 55)
Damage can resemble that caused by frit fly, the wireworms entering the base of the plant, tunnelling up the shoot and feeding on the young rolled leaves.

Chafer beetles (see page 57)

Aphids and leafhoppers (see page 60)
The bluish cereal leaf aphid, *Rhopalosiphum maidis*, is often found on the leaves of maize but its economic importance is not known under British conditions. Large populations of bird-cherry aphids (*Rhopalosiphum padi*) can be found on the leaves and cobs resulting in sooty moulds which develop on the sticky honeydew.

Thrips (see page 65)
Flecking of the leaves is typical of feeding activities by adult and immature thrips.

MITES

Two-spotted spider mite (*Tetranychus urticae*)
This tiny mite, only just visible to the naked eye, is more commonly found in Great Britain on plants grown under glass than on outdoor crops. In southern counties of England it occasionally infests maize, when the leaves become spotted with white marks and the foliage may be covered with fine web-like strands produced by the mites. Control measures are probably unnecessary.

MILLEPEDES (see page 66)

SLUGS (see page 68)

NEMATODES

Root-lesion nematodes
Maize can be severely damaged by the nematodes *Pratylenchus pinguicaudatus* and *P. fallax*, both of which attack the roots and

produce dark patches of dead tissue (necrotic lesions). The roots may later become secondarily infected with fungi or bacteria.

Cereal cyst nematode

Until recently, maize was thought to be immune to attacks by *Heterodera avenae*, probably because of the absence of white cysts developing on the roots of crops grown in infested soil. In any event, maize was considered a useful 'break' crop on light-textured soils such as the chalk downlands where *H. avenae* numbers are often high enough to affect intensive cereal cropping.

It is now known that larvae of cereal cyst nematode readily invade maize roots and cause serious loss of yield. Female development is impeded and white cysts very rarely appear on the root surface as in other infested cereals.

Stem nematode (*Ditylenchus dipsaci*)

No cases of stem nematode attacking maize have been confirmed in the British Isles. On the Continent, serious damage has occurred and in Germany marked differences in varietal susceptibility have been noted. Thus in field tests only 13 per cent of Inrakorn plants were infested, compared with 52–72 per cent for six other cultivars which yielded 18–31 per cent less than Inrakorn. It is not certain whether or not the oat race, so widespread in Britain (page 226), is responsible for the European attacks.

The position is thus similar to the one already described for rye (page 247). Both rye and maize are attacked by stem nematode in continental Europe. Any large-scale expansion of the area of either crop in Britain may allow one or more of our stem nematode races to cause trouble.

BIRDS (see page 75)

Birds, especially rooks, play havoc with seedling maize and, later on, with backward or lodged crops. For small areas, stringing with wires or black thread has given useful results. On larger fields, gun patrols and scaring devices have proved only partially effective in limiting bird damage.

MAMMALS (see page 77)

Chapter 11

MAIZE DISEASES

PLANT DAMAGE SYMPTOMS

Disorder	*Cause*	*Symptom*	*Page*
Damping off	*Pythium* spp.	Seedlings die before or soon after emergence.	258
Take-all	*Gaeumannomyces graminis*	Roots attacked but no symptoms on above-ground parts.	258
Smut	*Ustilago maydis*	Very large black galls on above-ground parts.	258
Stalk rot	*Fusarium* spp.	Rot at base of stalk; plants wilt, lodge; mainly after flowering.	259

Only a few diseases have been noticed in the maize crop which has been grown on a very limited scale in Britain.

Damping-off (*Pythium* spp.)
These fungi, which also cause a root rot in wheat (page 85) and other cereals, can cause pre-emergence or post-emergence death of seedlings, especially when soils are cold and wet. The damage can be reduced by seed treatment with thiram alone or as a combined dressing with gamma-HCH. Organo-mercury seed treatments should not be used because there is some evidence that they may damage seedlings.

Take-all (*Gaeumannomyces graminis*, syn. *Ophiobolus graminis*)
Maize has been shown to become infected by this fungus to an extent which throws some doubt on its value as a 'break' crop for wheat and barley. However, take-all appears to be unimportant as a cause of damage in maize. (A full account of the disease on wheat is on page 87).

Maize smut (*Ustilago maydis*)
This dieease occurs wherever maize is grown. It produces galls on any of the above-ground parts of the plant. These galls are often

large and spectacular, particularly when the cob is attacked. The fungus survives in the soil in the form of spores and in areas where maize is cropped frequently (more than about one year in three) the disease can be a serious problem. The fungus may be carried on seed though infection of seedlings directly from contaminated seed does not occur. However, the disease may be introduced to previously unaffected soils in this way. Seed disinfection will not control the disease which can only be achieved by adequate rotation. Smut is recorded every year in Britain, but at the present time it is a disease more of interest than of economic significance.

Survival in soil is important in the traditional maize-growing parts of the world, but its significance in Britain, at the northern climatic limit for maize growing, is not known. However, there is potential for silage maize in this country and if grain maize were to become more popular, the damage potential of smut would have to be considered in relation to cropping sequences.

Although successive crops of maize have been grown in Britain, in the absence of information on the ability of smut spores to survive in the soil in our climate, it would be unwise to grow the crop more frequently than one year in four.

Stalk rot (*Gibberella zeae* [*Fusarium graminearum*], *Gibberella avenaceae* [*Fusarium avenaceum*] and *Fusarium culmorum*)
Stalk rot, in which the base of the plant is attacked causing premature death and often lodging of late-harvested crops, can be caused by several fungi but in Britain the most important are *Fusarium* spp. The same fungi are able to attack other cereals (see pages 95, 176, 232, 250) and attention has already been drawn to some of the possible implications of an expanding maize area as a 'break' crop for winter wheat. The importance of maize as a carrier of indigenous cereal diseases in Britain is not yet clear. In France it is known to cause a build up of soil-borne inoculum which then causes an increase in Fusarium diseases of a following wheat crop. Perhaps the greatest danger lies in possible increase in the scab fungus, *Gibberella zeae* (see page 144), one of the causes of stalk rot.

The symptoms, which can appear very suddenly in a crop, consist of wilting, with the foliage turning a dull grey-green colour. This most commonly appears a few weeks after flowering as a result of early infection from the soil (later infections may go largely unnoticed). At this stage, in addition to the obvious foot rot condition, if the stem is cut lengthwise the pith is seen to be discoloured brownish pink in the lower parts of the stem, the discolouration sometimes extending far up the stem. Plants in this condition are liable to lodge in heavy rain or storms and this liability increases the later the crop is harvested. Thus grain maize tends to suffer more lodging than that harvested earlier for either silage or 'corn on the cob'. Varieties demanding high plant densities for high grain yield are more prone to the disease.

It is difficult to suggest useful control measures because there is still little experience of the crop under British conditions. There are distinct differences in varietal susceptibility. Delay at harvest should be avoided since it increases the risk of lodging following *Fusarium* infection. However, not all lodging is caused by *Fusarium* to which some varieties which lodge most easily are not particularly susceptible.

INDEX

Main reference page numbers are shown in **bold** type

FARMING PRESS BOOKS